Careers as an Electrician

By

Elizabeth Stewart Lytle

The Rosen Publishing Group, Inc.

New York

Published in 1993, 1996, 1999 by The Rosen Publishing Group, Inc.
29 East 21st Street, New York, NY 10010

Cover Photo by © Berle Cherney/UNIPHOTO Picture Agency

Revised Edition 1999

Library of Congress Cataloging-in-Publication Data

Lytle, Elizabeth Stewart.
Careers as an electrician / by Elizabeth Stewart Lytle.
p. cm.
Includes bibliographical references and index.
Summary: Describes the steps involved in becoming an electrician, from apprenticeship to full-time electrician work.
ISBN 0-8239-2885-3
1. Electricians—Vocational guidance—Juvenile literature.
[1. Electricians—Vocational guidance. 2. Vocational guidance.]
I. Title.
TK159.L98 1993
621.319'24'023—dc20 93-12776
CIP
AC

Manufactured in the United States of America

For my son,
Christian Ross Lytle,
who has,
at times,
proved to be a real "live wire."

About the Author

Elizabeth Stewart Lytle is a photojournalist, teacher, and communications consultant in western Pennsylvania. She shares an interest in construction trades and housing issues with her husband, a designer and builder. Lytle is the author of *Careers in the Construction Trades* and *Careers in Plumbing, Heating, and Cooling,* both published by The Rosen Publishing Group, Inc.

Mrs. Lytle is a contributing editor at two home-improvement magazines, *Your Home* and *Indoors & Out,* and she also writes articles for travel magazines. With nearly 400 magazine article sales to her credit, Mrs. Lytle's byline appears regularly in magazines published in the United States and Canada. In pursuit of travel adventure stories, she has flown a soaring plane and has twice sailed the Saguenay River Fjord region of Quebec Province, Canada, to observe and photograph the summer migration of whales.

Mrs. Lytle has received several writing awards as a newspaper journalist and as a freelance magazine writer. She also served three terms as a state director of the Pennsylvania Women's Press Association.

She has done considerable work in the Writer's Workshop at the University of Iowa, where she earned a degree in liberal studies and is a member of Alpha Sigma Lambda national honorary society. She is currently working on a master's degree in communications.

In 1991, the Franklin Area School District nominated Mrs. Lytle for the Sallie Mae First-Year Teacher Award, a program recognizing 100 of the nation's most outstanding new educators.

She and her husband have one son, Christian, who is also making a career in the construction industry.

Acknowledgments

Thanks are due to many generous individuals and organizations for their help in completing this book. Particular thanks go to William J. Stephens, producer of Audio/Visual Communications at Pennsylvania Electric Company, Johnstown, and to Mary Ann Van Meter of the *IBEW Journal* for photo support. Also thanks to the craftsmen and craftswomen who shared their viewpoints and their workdays; to the National Association of Home Builders and its educational arm, the Home Builders Institute; to the International Brotherhood of Electrical Workers; and to the U.S. Department of Labor Women's Bureau and Employment and Training Administration.

Contents

Introduction 1

1. Electricity: Technology's Lifeline 3

2. History of the Career 9

3. What Are the Job Options? 19

4. The Role of Labor Unions 37

5. Broadening Horizons for Women and Minorities 46

6. Apprenticeships: Learning by Doing 59

7. Other Routes to Training 80

8. Preparing for the Professional World 87

9. Opportunities to Specialize 101

10. Becoming Your Own Boss 108

11. A Day in the Life of an Electrical Worker 114

12. Jobs of the Future 127

Glossary 133

Appendix 137

For Further Reading 143

Index 145

Introduction

A clap of thunder shakes the walls. Lightning stabs the predawn sky, flickering with an eerie glow. Each jagged strike unleashes millions of volts of raw electrical power. Nature is staging a spectacular sound and light show, and I stand transfixed at the window until an unnerving crash rattles the glass pane beneath my fingertips. In the same instant, the glowing dial of the clock winks out. For the second time in less than a year, the electricity is out. I think of the transformer bolted to a utility pole up the street, wondering if it has received a direct hit. Until the emergency crew members arrive in their bright yellow ladder truck, don safety gear, and put their skills to the test, my neighborhood will be dragged backward into a way of life left behind a century ago. I wonder how long the power outage will last. An hour? All morning? A full day? But I have underestimated the resourcefulness of the electric company, and twenty minutes later the invisible stream of energy begins to flow again, its presence indicated by the soft whir and hum of household appliances restored to working order.

I am able to keep my vow to be at the computer early this morning, working on the book you are reading right now. Electricity has revolutionized the way people live and work. It has transformed and simplified our tasks, giving us leisure time and lighting our world around the clock, if we so choose. And now, after nearly a century of the machine age, electricity has merged with electronics to bring human innovation to yet another level.

Mass communication, data transfer, medical science, artificial intelligence, global transportation—all of these depend on electricity. As we approach the end of the twentieth century, perhaps many of us have begun to take electrical power for granted. After all, it has been more than a century since our nation was first transformed by abundant and affordable electric power—first the cities, then the towns, and finally even the isolated farms and ranches. Perhaps it is also time to consider the millions of trained craftworkers who have learned to harness the incredible power of electricity. Certainly it is time to encourage your generation to discover and direct this mysterious energy. By studying theories of electricity and their practical applications, you can join more than one million craftworkers who are busy installing and maintaining sophisticated electrical systems and equipment. Their work commands respect and brings them a steady income and good fringe benefits. Electricians and electrical workers enjoy mobility, for they can take their skills anywhere in the nation and quickly find work. The field offers a variety of careers to choose from, such as electric utility, construction, and maintenance. The electrician's basic training provides an excellent basis for specialized training. Modern communications, electronics, lasers, and robotics are just a few of the emerging fields that bring together technology and electricity. Electricians even have a role to play in space exploration, so—quite literally—the sky's the limit for electricians in the twenty-first century.

1

Electricity: Technology's Lifeline

Nearly one million people earn their living in the electrical trades, with the demand for skilled craftworkers continuing to grow. More than half of these jobs fall into the construction and electrical mechanic/repairperson categories. Journeyworker electricians are among the highest paid of all union construction workers, averaging $34 per hour in 1998, according to the Construction Labor Research Council. While construction electricians are the best paid in the industry, the number of hours they work is determined by weather conditions. Those trained in electromechanics or electrical repairs earn $18 to $25 per hour year-round. Most electrical workers generally receive fringe benefits, which boost earnings packages. Typically these benefits include paid vacation, retirement benefits, and health and life insurance plans, all of which are discussed further in the next chapter.

Because jobs in the electrical industry have a common base in theory and applications, it is possible for workers trained in one job category to shift to another career within the industry. For example, a qualified journeyworker may switch from the construction field to a career in electrical maintenance or power plant work, or vice versa. Few careers offer such mobility.

Nearly all jobs in electrical work are open to people without a college education. Skills training involves a combination of technical instruction and supervised on-the-job learning over a span of three to four years. This type of training provides the instruction an electrician or electrical worker needs to enter the craft. Such an apprenticeship prepares you for interesting, well-paid work. You will be paid while you learn, receiving regular raises as your skills increase.

As electrical workers receive thorough theoretical training, they must never forget the potential for danger in their work. From their earliest days as an apprentice, it is important that they learn to treat electric current with respect. Doing careful work becomes a deeply engrained habit, for these workers know that their own personal safety, as well as the well-being of others, depends on that expertise.

RECRUITING EFFORTS

Through active recruiting efforts, construction and utility industries are working to attract qualified female and minority candidates as apprentice electricians. The shortage of skilled workers that is predicted for the early twenty-first century means employers will have to compete for those workers. The best-trained, most experienced electrical workers will likely have their choice of job offers anywhere in the country. The next decade promises to be an exciting time to begin a career in the electrical field.

Electrical workers harness the awesome power of electricity, commanding it to heat, cool, and light the cities of the world and to power the machines and perform the work that have transformed human civilization.

With training and experience, you could be one of these workers.

The Bureau of Labor Statistics of the U. S. Department of Labor estimates there are now more than 500,000 journeyworker electricians in the United States. More than 30,000 of them are apprentice electricians. The Bureau also predicts there will be nearly 6,500 job openings annually for construction electricians alone through the year 2000. In the utility group, the need for electrical power line installers is expected to keep pace with job growth for all industries. The same is true for maintenance electricians. In addition, electricians who can apply their training to the related field of electronics will find almost limitless opportunities in laser and fiber optics technology, robotics, space, defense, and medical research.

The population of the United States is expected to continue to grow significantly. By the turn of the century, approximately 268 million people will live in the United States. These people will need places to live, businesses, stores, factories, hospitals and medical care, schools, and community services. All of these aspects of modern life involve work for the construction electricians who will wire the buildings and for the electrical workers who will build, install, maintain, and repair the electrical systems and equipment needed within these structures.

REAL JOBS FOR REAL PEOPLE

Let's examine the three branches of electrical careers. A good place to start is with the utility branch, for it is here that electrical power is first produced or generated.

Utility Branch

Electrical utilities employ skilled workers to operate generators and to distribute power via overhead transmission lines throughout the utility service area. There are also jobs for maintenance electricians, line workers, troubleshooters, and cable splicers. One or more of these electric utility companies serves every state in the nation, providing thousands of jobs nationwide. Since a utility company's service area may be more than 100 miles wide, a network of substations and regional maintenance centers is established to provide a quick response for emergency repairs and regular customer service. These facilities also employ electricians in the utility categories.

Construction Electricians

Construction electricians make up an essential part of any building crew. The construction electrician's work is categorized as either "rough work" or "finish work." Regardless of the type of building under construction, electricians work with wire, conduit, and current. They install, connect, and test electrical wiring systems throughout the structure; provide power hookups for heating and air conditioning systems; and, in residential work, install hot-water tanks, automatic dishwashers, and clothes dryers.

Construction wages have traditionally led the three divisions, but such rates may not be steadily available throughout the year. There are also differences in wages because of geography.

Maintenance Electricians

Maintenance electricians keep the electrical equipment in factories, offices, and residential buildings working.

Blueprints and diagrams are part of their work, too, and they may use meters and other test equipment to locate electrical problems.

The only real difference between the contruction electrician and the maintenance electrician is in the application of skills for each setting. Whereas the emphasis in construction is on installing and testing new systems, the maintenance electrician focuses on regular maintenance to keep systems operational, by repairing or replacing equipment that fails.

Manufacturing firms and other companies that are open twenty-four hours per day may employ three maintenance crews, one crew for each eight-hour work shift. The day shift is usually responsible for routine maintenance; electricians on the afternoon and night shifts concentrate on preventing costly production losses. If there is an emergency, these skilled workers may be expected to advise management of the different options available, such as shutting down equipment for repairs.

Special Fields

There are several specialized areas of the electrical industry, notably the electric motor shop industry, the electric sign industry, marine and shipboard electrical services, cable television, and, increasingly, the alliance between electricity and electronics. Fiber optics, lasers, and robotics are among the new technologies that forge and encourage a link between electricians and electronics workers. Two industries, communications and manufacturing, have also helped to strengthen the link between technology and the world of the electrical worker. These industries are discussed further in a later chapter.

STABLE, REWARDING CAREERS

Today's electricians and electrical workers enjoy stable employment and are among this country's highest-paid skilled workers. As with other occupations involving the use of power tools and electricity, there is an element of danger. These workers have been specifically trained to handle dangerous situations. All electricians learn to follow National Electrical Code specifications, in addition to any state and local electrical guidelines that apply.

Electricians and electrical workers earn a great deal of respect because of the type of work they do. Their training and experience are rated highly in the work world.

2

History of the Career

Humankind has spent centuries trying to unravel the mysteries of electrical power. Although we do not completely understand it, we have managed to harness this immense source of energy. Electricity is the name we use for something no one has actually seen, and whose real nature remains mysterious. The zigzag flash of light that appears in the dark sky during an electrical storm is not electricity, but rather an effect of it. We speak about the production and the generation of electricity, but that is somewhat misleading, since electricity really cannot be created or manufactured. What we can do is gather and harness some of the endless supply of electricity that already exists in the world.

For many years, electricity existed as a sort of scientific toy. Lacking the knowledge to put electricity to work, scientists could only guess at its potential. No one had yet determined the rules or laws that electricity always follows. No one knew where this intriguing energy source "hid." In fact, it seemed to originate in many unlikely locations. The entire world appeared to be an electricity reservoir, permitting anyone with the knowledge and the resources to gather electricity in useful amounts. Magnets, chemicals, and friction are the typical tools used to gather electricity. The form of electricity used for lighting homes, powering urban transit systems, and running factories is produced by magnetism and moves in a current, like water through a pipe.

Although electricity makes possible many kinds of mechanized work, its ability to produce artificial light is probably the feature that has had the greatest impact on human life. From the dawn of civilization, people have valued light as a source of warmth, illumination, and security. Since fire was discovered, humans have labored to control the power of light. Over 6,000 years of history, despite advances in basic science, no real progress occurred in the development of the lamp until well into the nineteenth century. Some city dwellers were able to use the natural-gas lamp, but most people continued to rely on oil lamps and lanterns to illuminate their homes and workplaces. It was not until 1879 that Thomas Alva Edison unlocked the secret of electric light.

Edison applied his genius to many areas of electricity, and his early work inspired other inventors, prompting a steady flow of improvements to the incandescent light. At about the same time, Edison teamed up with another inventor, Alexander Graham Bell, to refine Bell's invention of an instrument that used electricity to transmit the human voice over a wire.

Once scientists understood the nature of electricity, they learned to manipulate its immense power to do their bidding. Inventions and devices previously deemed unworkable fulfilled their promise when powered by electric current. Earlier useful inventions were "electrified" to work even more efficiently. A new era had begun.

The American Centennial Exposition of 1876 is generally accepted as the beginning of the age of electricity. Among the electrical products showcased were storage batteries, electric bells, and telegraph equipment. Bell exhibited his new invention, the telephone.

Shortly after that, Edison opened a factory that produced direct-current generators. Electric streetcars were next, and within ten years the Westinghouse Corporation built America's first successful power station, which generated alternating current.

Of the fifteen industries created since 1870, only automobile manufacturing has created more jobs than the manufacture of electrical machinery and supplies. A related industry, electric welding, triggered growth in both the construction and transportation industries. Electric motors replaced water and steam power in many industries. Electric current illuminated factories, schools, hospitals, and homes. It also powered labor-saving devices and appliances, raising everyone's standard of living. In September 1882, Edison launched a new lighting system that illuminated one square mile of New York City. Ventures such as this helped the world to acknowledge that a remarkable new industry had been born. In the years that followed, the U.S. government encouraged rural areas to electrify, easing the burden on farmers and improving life for rural residents. Before long, electric power lines served every corner of the nation.

Jobs in an Infant Industry

The first electrical workers built a network of power distribution lines, resulting in a the job description "lineman." Today's utility company service personnel include male and female workers. In the early days this work was exciting and highly dangerous. The fatality rate was high, and workers found it nearly impossible to obtain life insurance. They worked long, hard hours for relatively low pay.

As electrical power became more readily available, demand developed for workers who could wire houses, buildings, and commercial establishments. These well-trained people had to be knowledgeable about circuits and equipment. They did most of their work indoors and became known as "inside" electricians. As generating capacity grew, industries converted to electrical power and boosted production capacity, creating even more jobs. The price of goods and services fell, and workers with regular cash wages could afford to buy them. Next came the need to service, repair, and maintain electrical systems and equipment, and—you guessed it—even more new jobs.

Electric motor repair and electric sign service emerged as distinct parts of the industry, but overall the pace of change slowed until the mid-1960s, when electronics burst upon the technological scene. A second revolution in the use of electrical power and systems is upon us. Like the first, it promises to usher in new products, processes, and jobs.

THE MODERN-DAY INDUSTRY

If you find the subject of electricity and its many applications of interest, the exciting and rewarding field of electrical trades may be the right career choice for you.

Although most craftworkers learn their skills in an apprenticeship, a college degree is the route for job opportunities in research and development, electrical engineering, robotics, manufacturing technology, systems analysis, and teaching at all levels, from vocational-technical high school to university engineering departments.

Technical and College Degrees

As technology advances, the individual with a post-high school technical degree will be the most sought-after worker in the electrical field. Two-year programs teach skills in building, operating, and maintaining electrical equipment including the motors and controllers used in manufacturing, plus the generating and distribution apparatus used to produce electricity. For those interested in engineering, a four-year degree program in electrical engineering will prepare you to design systems.

A DAY ON THE JOB

Alan, a journeyworker electrician, is on a house call to a suburban neighborhood. A mixture of carefully maintained split-level and ranch-style homes stand well back from the street on tree-shaded lots. The address on the work order matches that of a ranch with an obviously new patio linking the residence with its garage. As part of this home-improvement project, the property owner wants to have two safe outdoor electrical outlets installed in the exterior wall.

"The way I understand it," Alan explains, "these folks have been using a heavy-duty extension cord plugged into an outlet in the garage to power a weed trimmer in the yard and small appliances on the patio. Now they realize it's time to do things the right way. That means installing a couple of ground fault circuit interrupter (GFCI) outlets. This is absolutely the way to go, since GFCI outlets protect against electrical shock outdoors and in areas that could be damp enough to conduct electric current. The brand we use shuts down in one-fortieth of a second."

The contracting firm's logo, a silver lightning bolt, is visible on the side of the van as we pull up at the curb, and Alan wears a dark blue jacket with the company logo. Nevertheless, he has his photo ID card ready to show to the client, who answers the door.

After a quick look at the project site, the electrician agrees that the job should be relatively simple. He will be able to draw power for one of the new outlets from an outlet on the interior side of the wall, located between the same pair of wall studs. For the other, he can tap into an electric junction box in the basement. The latter method involves fishing some wires through the wall, but it allows placing the second outlet where the client wants it.

None of these outlets is expected to be put to heavy use in operating large power tools or appliances that operate at more than 300 watts. Had the client needed that much wattage, Alan later explained, he would have suggested installing a new circuit.

The first step in this new job is to shut off power to the outlet the electrician will be working on. This is done at the main electric panel in the basement. Each of the circuit breakers is labeled, indicating the system by which electricity is directed to the heating and cooling systems, the laundry and kitchen appliances, and to wall outlets and light fixtures in individual rooms.

"You want to be sure not to overload the circuit you branch out from," Alan says. "Unfortunately, there's no simple method for figuring this out. A rule of thumb I learned during apprenticeship is not to power more than eight outlets and lights on a simple 15-amp circuit, or more than ten outlets and fixtures on a 20-amp circuit.

"We need a 20-cubic-inch outlet box to run the new wire from, which means replacing the existing one with a 'remodeling'-type box. That's part of the regular parts inventory I carry in the van, along with boxes of residential and commercial-type wiring for both interior and exterior uses. Everything we stock is U.L.-listed; that means the material has been tested and approved as safe by the Underwriters Laboratories.

"Cutting into walls and fishing wires through them can be a tricky business. You have to be very careful not to nick other wires or pipes that may be hidden from view. That's why these retrofit jobs take so much longer than wiring a building during the original construction phase."

Cutting the hole for the new outlet box requires careful planning and a saw—choices include a jigsaw, a keyhole saw, or Alan's personal favorite, the versatile and powerful reciprocating saw.

"To decide where to place the outside outlet, we need to make sure we're working between the same pair of studs as the inside outlet. I use the center of a window or door as a reference."

The journeyworker expertly fishes the wire from the new outlet box hole into an entrance space that he explains is called a "knockout hole" on the interior electrical box. With a few deft twists of his wrist, he unscrews and removes the interior outlet, thus gaining working space.

To wire the interior outlet, he uses a hand tool to

strip three quarters of an inch of insulation from the ends of the white and black wires. Bunching together the new and existing white wires, with a five-inch pigtail, Alan secures them with a WireNut. Next, he wraps the pigtail around the silver or "white" designated screw. He does the same with the black wires but secures the pigtail to the opposite side of the outlet. After securing all the bare ground wires with a pigtail and a WireNut, he fastens the free end of the pigtail to the ground screw and pronounces this portion of the job completed.

Alan moves outdoors to wire the exterior GFCI outlet. He begins by removing twelve inches of the sheath from the cable and sliding it into the box. Next, the box is fastened securely to the siding with screws. He secures the ground wire to the metal box, then makes the other necessary wiring connections.

"When the exterior of a house is brick, masonry, or stucco finish, you can surface-mount a special exterior electrical box," he notes, "but working with this narrower lap siding, we'll have to use GFCI outlets and covers designed for horizontal installation."

After caulking between the metal box and siding to prevent air infiltration, the electrician installs the gasket and waterproof cover.

In a matter of minutes, electric power is restored, and he checks the GFCI by pressing the test button. "I saw this installation performed a dozen times before I actually carried out the steps myself," he explains while packing up his tools. "The first few times we watched on videotape in apprenticeship classes. Then I watched as the journeyworker who trained me went through the steps, explaining each one as he went along. Finally, when I could explain the process orally, I got to wire my first outlets, under the watchful eye of my supervisor.

"I've probably done a hundred of these since, but no matter how many you do, it still takes the electrical inspector to say you've got everything right."

Afternoon finds Alan in the warehouse of the electrical contracting firm for which he works, loading materials for a house-wiring job. For the first several weeks since groundbreaking, a portable generator has provided the electricity to run power tools. Now that the basic shell is erected and roofed, it is time for the rough wiring to be undertaken.

"Let's see, we need a 200-amp entry panel with room for twenty-four circuits; electrical house wire in 250-foot lengths, for both exterior and interior use; a couple dozen 4-inch steel work boxes with knockouts and covers; and duplex grounded receptacles and switches. Of course, we'll check this materials list against the electrical drawings to be sure we've got everything," he adds. "And we'll be doing some more GFCI outlets—in the kitchen and both bathrooms for sure.

"There will be overhead light fixtures for every room and the hall, plus outlets galore. We'll be wiring both a home entertainment system and an integral security system with room-to-room intercom and stereo. This is the kind of project I really enjoy. Starting from scratch with sophisticated materials and taking responsibility for the entire job, from the beginning clear through to the electrical inspector's visit.

"Planning the lighting requirements for a new house often gets short shrift, considering that it comes rather late in the design process, often when the budget is tight. Not in this case, though. The owner is an ophthalmologist who knows the importance of good lighting from a health standpoint as well as an environmental viewpoint." Tapping the roll of architectural drawings in his hand, Alan adds, "These plans show that a lot of thought went into the lighting design of this house. There are recessed and surface-mounted fixtures, both incandescent and fluorescent, interior and exterior. Some of the exterior lighting is actually solar-powered. Other elements contain motion- or heat-detecting devices, or both, and switch on automatically when their sensors are triggered. This house has more than a dozen light fixtures in the kitchen alone, taking into account the need for ambient light, task lighting, and general illumination. Along with the track lighting and ceiling fans you find in most new homes, these plans call for some purely artistic lighting to show off some artwork and houseplants.

"But I'm getting ahead of myself. All that is the finish wiring, the icing on the cake. Today we get started with the nitty-gritty, what they call 'roughing-in.' It will be a long day of drilling holes through studs and pulling wire. We may have some conduit to bend, too, and we'll wire up the panel box."

It is obvious that Alan looks forward to the job, and from what he has told us, he will be amply rewarded for his efforts.

3

What Are the Job Options?

If you like variety in your daily work, the electrician's career is a good choice. Depending on the specific area you choose, an electrician's job can be adventurous, exciting, or downright dangerous. It can also involve travel along the nation's waterways or to distant ports of call.

You might work for an electric utility company, climbing to dizzying heights and working on high-voltage lines, or you could be an instrumentation technician, hopping jet flights on short notice to remote locations where you alone have the skills to put important factory equipment back on line.

Or you might choose to work closer to home, erecting and installing electric signs at commercial locations, wiring new houses, or installing and maintaining electrical equipment. As a journeyworker, you might work in such varied environments as residential, commercial, or industrial construction projects, for an electrical service contractor, on a manufacturing plant's maintenance crew, or for a shipbuilding company. With skill and experience, you might one day supervise a crew of electricians in one of these locations, or even your own electrical contracting business. Regardless of your choice, variety can still "spark" your work life.

Because electricians are in demand throughout the nation, you will have a nearly unlimited choice about where to live and work. Should you choose to become a marine electrician, the Great Lakes, both coasts, and the Gulf of Mexico beckon. Construction projects dot the national landscape, and there are likewise many opportunities for electric sign builders, power station workers, and people who can build and repair electric motors. Let's take a closer look at what a journeyworker does in each of these specific occupations.

CONSTRUCTION ELECTRICIAN

Inside construction electricians lay out, assemble, install, and test electrical circuits, fixtures, appliances, equipment, and machinery. Among the electrical systems they are expected to understand are lighting, heating, cooling, and control systems in various types of structures. A quick review of the types of buildings in your neighborhood—homes, schools, shopping and office complexes, hospitals, and factories—makes apparent the diversity of services required.

At new homesites, electricians install electric circuits, appliances, and equipment including water heaters, smoke or burglar alarm systems, intercoms, and heating systems. Commercial and small industrial building projects involve electrical systems on a larger scale, for instance, wiring the heating and cooling system for a supermarket or an office complex. Here you are likely to find security and fire protection systems on a larger scale. Another category of electrical construction is industrial or "large" work. Once you know that this category includes new factories, petroleum or chemical plants, and other production facilities, you can understand how the term "large" applies. Such jobs can last a year or longer.

To do their work, construction electricians follow blueprints and architectural specifications detailing the wiring systems. They could be involved in installing electrical machinery, controls, signal systems, and communications equipment. Electronic controls are increasingly used in homes and commercial buildings, requiring construction electricians to learn about this new aspect of wiring.

The primary employer for the nation's construction electricians is an electrical contractor, and there are about 50,000 such firms in the United States. Qualified electricians can find work anywhere, but the largest concentration is in indutrialized areas and in cities. Electricians are among the highest paid of all construction workers. In January 1998, union electricians on construction projects in the United States averaged $33.90 per hour, an increase of nearly $5 per hour from the rate paid five years earlier. These high wages are offset, however, by periods of little or no work. To help construction electricians deal with an erratic seasonal work schedule, supplemental unemployment benefits (SUB) and severance pay allowances help smooth out the financial highs and lows. These two benefits give a measure of security at such critical times as layoffs and job terminations. A growing number of electrical workers have such protection. Excellent health insurance coverage is typically provided as a fringe benefit.

MAINTENANCE ELECTRICIAN

The basic difference between a construction electrician and a maintenance electrician is how each applies the skills of the craft. Maintenance electricians are responsible for maintaining existing electrical systems, equipment, and machinery. It is their job to find the source of trouble when a piece of electrical equipment fails to operate properly. To do this, the maintenance electrician uses a technique known as troubleshooting. A number of tools and testing devices are used, including voltmeters, voltage testers, test lamps, oscilloscopes, and wattmeters. Other tools include power factor meters, instrument transformers, logic probes, and digital devices.

Maintenance electricians may install lighting circuits, distribution and load centers, fuse panels, magnetic starters, safety switches, electric motor controllers, rectifiers, electrical machinery and equipment, and many other types of apparatus.

To service equipment such as motors, generators, controllers, and switch gears, the electrician cleans parts and makes minor repairs and adjustments according to the manufacturer's suggestions or accepted industry standards.

Preventive maintenance is stressed. Maintenance electricians periodically service equipment and systems, making necessary repairs to ensure continuing operation of the facility. These activities follow a specific timetable for machinery and equipment, with the electrician documenting what was done and when it was done in a series of brief written reports.

Because many companies operate twenty-four hours a day, they employ three shifts of maintenance electricians, each working eight hours per day. Generally, the first shift takes primary responsibility for routine maintenance, whereas the second and third shifts prevent costly production losses and interruptions. During emergencies, the electrical maintenance supervisor may be called upon to decide if an immediate shutdown is necessary. When this happens, the time needed for repairs must be estimated.

Necessary skills include being able to plan, assemble, install, and test electrical circuits and systems. In industrial applications, this may include complex sensory, monitoring, and other control devices. The electrician must interpret and work from blueprints and wiring diagrams and be familiar with the National Electrical Code.

In addition to the tools owned and used by the construction electrician, a maintenance electrician may also need open-end wrenches, a ball-peen hammer, punches, calipers, and a socket set. These items cost $300 or more.

One of the primary attractions of work as a maintenance electrician is stable employment. These people often devote their entire career to a single employer. Most of their work is performed indoors. Still, they must frequently work in close quarters and may be exposed to extremes of indoor temperatures.

Maintenance electricians generally enjoy their work. As a group, they are resourceful and inventive. And they are in demand; job opportunities are expected to increase faster than the average for other electrical occupations through the end of the decade.

ELECTRIC MOTOR SHOP JOURNEYWORKERS

These workers are highly trained, utilizing a variety of mechanical skills in addition to their knowledge of electrical concepts and theory. Machines often play a part in the work, which is done in an electric motor shop. Journeyworkers typically operate such devices as metal lathes, static and dynamic balancing equipment, drill presses, power saws, hydraulic presses, and coil-winding equipment. As you might imagine, shop training is part of this electrician's education.

Repair and servicing of electrical machinery and equipment may be done in the shop, or workers may make service calls. Typical in-shop projects include rebuilding electric motors, generators, starters, and controllers. Journeyworkers also rewind transformers, relays, magnetic brake coils, and other coil windings; they fabricate switchboards and panelboards. Such tasks require soldering and welding.

Those who work in electric motor shops are mechanically inclined, but they must also be skilled at analyzing equipment to diagnose the source of the problems. They must be able to read and understand complex control wiring diagrams and controller and starter wiring schematics.

Diagnostic tools include the voltmeter, ammeter, tachometer, megohmmeter, frequency meter, and wattmeter. This worker maintains a collection of tools worth about $350, which includes such items as a ball-peen hammer, rule, socket wrenches, calipers and micrometers, an assortment of punches and chisels, pocketknife, combination wrenches, pipe pliers, pipe wrenches, hacksaw, hollow-head wrenches, diagonal and side-cutting pliers, and many special winding tools.

While the work is interesting and may present challenges, it can also be physically demanding.

MARINE ELECTRICIAN

Ships and barges need lights and electricity, and it is the marine electrician who provides them. Almost every operation aboard modern ships requires electric power. Many ships have electric propulsion motors powered by generators, with complex wiring systems connecting the two. In addition, communications and navigation equipment, pumps and auxiliary machinery, steering gear motors, and other vital shipboard systems run on electric power.

A network of electric cables, often more than a mile in total length, transmits power where it is needed. This cable is installed and repaired by marine electricians.

Thousands of electricians are employed by marine repair and service companies with dock or harbor-front locations. Others are employed by shipping or transport companies as part of the working crew.

The key to keeping marine vessels "shipshape" is careful regular maintenance. Marine electricians spend much of their time on maintenance duties, where the work focuses on installing new systems and adapting existing ones. Here, too, resourcefulness is needed. Since an exact replacement for a piece of equipment may not be available when a ship is at sea, the electrician must analyze the situation, decide upon a course of action, and then carry it out, often adapting parts and materials to fit the need. As you can see, natural ability as a mechanic is a requirement. Another important duty is finding grounds and other faults in electrical systems.

The marine electrician must know some techniques unique to shipbuilding. Watertight seals are necessary throughout a vessel, and there are specialized methods for installing cables and fittings on board a ship. Marine electricians are also required to learn the U.S. Coast Guard regulations that apply to their work. Offshore drilling rigs need marine electricians to service and repair electric wiring and apparatus.

The marine electrician's toolbox compares to that of a maintenance electrician, plus some specialized tools—cable skinners to remove the protective armor from cables, and packing tools to pack watertight terminal glands.

The marine electrician is on the move through much of the workday, bending, kneeling, climbing, and standing. On occasion, it may be necessary to lift heavy weights. Good health and physical strength are essential. The work is done indoors and outdoors, and under variable weather conditions. A marine electrician may be on deck in freezing weather or laboring in a hot engine room.

Employment is generally stable, with ample opportunity for interesting and challenging assignments, sometimes with travel. Many of today's marine electricians received their training during military service.

Employment opportunities exist in coastal regions, along the Gulf Shore and the Great Lakes, and on major rivers. Job opportunities also exist in foreign countries.

UTILITIES AND POWER PLANTS

Thousands of trained workers are needed to operate and maintain the network of electric generating stations that serve the nation. While varied in size and capacity, these centers generate and distribute electric power around the clock, fulfilling a commitment of service to millions of customers.

Keeping the facility in top operating condition at all times requires a well-trained staff to install, service, and maintain the complex systems that deliver electric energy. The job classifications are many in the industry, but our discussion focuses on occupations requiring knowledge and skills parallel to those of other electrical workers.

But first, how is electricity produced? An electric generating plant is a fascinating place to visit; although electric current is being produced in tremendous amounts, there is nothing to be seen of it. Huge dynamos (also called generators) revolve at high speed day and night, transforming the energy of huge engines or steam or water turbines into invisible electric current, which is transmitted at lightning speed to cities and towns across a network of copper wires. Despite all that goes on in a generating room, it is very quiet. Giant dynamos turn at 3,000 revolutions per minute, yet the only sound is a gentle hum. The engines or turbines run quietly, too, although the power they generate as you look on may be sufficient to run the homes and factories of an entire state.

Controlling the speed of the dynamos is very important. Sometimes several generators feed into the same lines simultaneously, requiring an exact match in their alternations. It is highly economical to transmit current at a high voltage, and then "step it down" again at its destination.

Steam-driven power plants are usually built on the banks of a river, since they require ample supplies of water. So, too, are hydroelectric facilities. Nuclear fuel is certainly a possible source of energy production, although public concern over safety and the enormous costs of building the facilities have stymied this part of the industry.

Although the current trend in building power plants

favors larger, fully automated stations, a considerable number of workers is needed to maintain and repair machinery and related equipment.

POWER PLANT MAINTENANCE ELECTRICIAN

This electrician performs duties similar to those of a maintenance electrician, except that the power plant worker is concerned with equipment that generates electric energy. They use many of the same tools and generally work indoors, focusing on preventive measures that guard against major and costly breakdowns. Power plant electricians must be able to make quick, accurate decisions in carrying out assignments, especially when unforeseen circumstances combine to demand immediate action.

Power plant electricians are usually shift workers, since generating stations never shut down.

Once generated, electricity is instantly ready for use. Delivering it to the customers of a utility plant requires a complex network of electrical conductors, which are the responsibility of transmission and distribution departments. These units are staffed by workers with varied duties involving the building, maintenance, and repair of the system. Their work ranges from driving trucks and operating pole-setting equipment to managing complex wiring installations. Our focus is on the jobs lineworker, cable splicer, and troubleshooter.

Lineworkers

Lineworkers, who make up the largest single occupation in the utility industry, construct, set anchor, and install the guy-lines for utility poles and structures and string the high-tension lines. They must be familiar with pole line construction methods and the terminology associated with this work. They build transformer structures and install them, along with protective devices, connecting such equipment to power lines. For this reason, they need adequate knowledge of transformer theory and connections.

Lineworkers often work from aerial baskets, which are insulated "buckets" attached to a boom mounted on the bed of a truck. Such equipment makes it easier and safe to perform overhead work, since aerial baskets can usually be positioned with ease. In some power companies, lineworkers specialize in particular types of work. Some may be assigned to new construction while others perform repairs.

Their work is often done at extreme heights and under hazardous conditions. At times they work on energized power lines and equipment in foul weather. When wire, cables, or poles break, it means an emergency call for a line crew. Lineworkers are often the first on the scene following a hurricane, storm, or tornado, working to clear debris and fallen trees so that they can restore the electric service. Lineworkers are well trained in safety measures, for an element of danger exists in their work.

Troubleshooters

Often working alone on service calls, troubleshooters drive a service truck equipped with tools and materials necessary to handle routine calls, moving from one assignment to another on orders from a central office. Mobile telephone or two-way radio provides contact with headquarters.

At times this worker disconnects or isolates the

defective component or circuit from the rest of the system, allowing safe operation until a repair crew can correct the breakdown. Specific responses depend on the nature and extent of the breakdown, but a troubleshooter might also maintain overhead and underground primary lines and equipment; use hot line tools and rubber protective equipment in working on energized lines and apparatus; test, refuse, or replace transformers, fuse cutouts, lightning arresters, and similar devices to restore service; clear shorts or grounds from lines; restore downed wires and trim trees that create a hazard to wiring; make temporary repairs to poles, guys, and fixtures; report repairs requiring a line crew; clear trouble at automatic substations; inspect the condition of lines and distribution equipment; and perform routine clerical work including field and time reports and material requisitions.

Cable Splicers

These workers install and repair single and multiple conductor insulated cables on utility poles and towers, as well as those buried underground or carried via underground conduits. They make connections between electrical conductors, insulating the conductors and sealing cable joints with a lead sleeve.

Cable splicers know the various methods and techniques necessary to bond and ground lead sheaths of cables for all voltages. They also know how to weatherproof and fireproof cables.

Journeyworkers' duties include complicated construction and maintenance work on dead or energized underground services and conduits, breakers and fittings, service boxes, meter boxes, current and conduits, potential transformers, and motor and lighting circuits. Cable splicers must also find and repair faults in circuits and equipment. Cable splicers must be familiar with the characteristics of metals and insulating materials used in their work. A knowledge of conduit and duct work is also essential.

Cable splicers work in manholes, in vaults, and on poles when performing their duties, a substantial amount of which pertain to the installation and maintenance of underground lines and overhead cables and the changing of cable systems layouts. They also make all types of polyphase transformer installations, perform phasing and phase rotation tests on polyphase circuits, and when required, trace and clear grounds and open circuits on cable systems. Occasionally, work in this job category is interrupted or delayed, and the cable splicer may be assigned to other duties in the division.

At times all three of these jobs—lineworker, troubleshooter, and cable splicer—require working from detailed instructions and blueprints. Good judgment and independent thinking are required to come up with solutions to problems. These jobs require patience, precision, manual dexterity, and vigilant safety consciousness. Initiative and ingenuity are also essential.

Experienced journeyworkers are often assigned to supervise other workers on their crew. Lineworkers, troubleshooters, and cable splicers must be familiar with the National Electrical Code, which is a consensus code covering the construction, repair, and maintenance of overhead and underground electrical conductors and equipment.

The work is usually performed outdoors, sometimes

under challenging conditions, such as following a hurricane, storm, or tornado when communities are without electric power.

Because these jobs require workers to climb, bend, and work in awkward positions, physical stamina is required. Good visual perception is also important.

Journeyworkers employed on substation work provide their own tools, typically a knife, hammer, pliers, hacksaw, 14-inch pipe wrench, large and small screwdrivers, combination square, level, adjustable and box-end wrenches, and a chisel and punch set. Those performing line work usually furnish climbers, safety belts, pliers, hammers, wrench and hand connectors, wooden rules, screwdrivers, adjustable-end wrenches, lineman's pliers, and wire-skinning knives.

Those who perform utility line construction and maintenance describe their jobs as interesting and sometimes adventurous. Certainly, these electricians are knowledgeable and highly skilled. When a natural disaster overwhelms the resources of a single utility company, line repair crews of utilities in neighboring regions and states regularly pitch in to help, working long hours far from home and proving that dedication is another attribute of their profession.

Employment for electric power line installers is expected to keep pace with a growing and increasingly mobile U.S. population. However, cable splicers and lineworkers in the cable television and telephone industries will see a small decline in job totals, as a result of increased system efficiency and new technology that is less vulnerable to weather outage. Lineworkers employed by utility companies and construction firms will enjoy the most job security.

Training

On-the-job training is a continuing factor for power company employees, with field demonstrations and classroom sessions continuing over the course of a career. This enables employees to qualify for more difficult assignments, in addition to keeping current with technology.

In addition to employer-provided training, line and cable workers may be sent to short-term "schools" sponsored by the manufacturers of cable installation equipment. At times these manufacturers send their trainers to the job site to demonstrate new techniques and equipment.

New teaching methods are being used to augment or even replace traditional classroom methods. Videotapes, films, computer-assisted instruction, and programmed learning materials are among the most effective methods. The use of learning labs has increased also. Equipment such as utility poles, cable-supporting clamps, and similar fixtures simulate an actual work environment. Those in training practice working on poles while keeping their hands free. As an example of this teaching method, one exercise involves a game of "catch" played with a basketball while the trainees stand on the poles. The emphasis, however, is on worker safety at all times.

Electric Sign Service Jobs

Electric signs have become so commonplace that communities of all sizes now have them. Whether of neon or illuminated panel design, they are electric-powered and require the skills of a trained electrician to build, install, maintain, and repair.

The electrician who installs signs first plans a job layout from detailed drawings that illustrate construction design and the materials to be used. The work requires knowledge of proper methods for hanging signs and using rigging. Because the signs are often large and heavy, they are installed in such a way as to withstand strong winds. The installer must understand applied physics, as well as various types of supports and their safe working loads.

Electric sign installers are also metalworkers. They bend sheet metal and plastic into various forms, cut out metal and plastic parts, and drill or punch holes. Parts are fastened by riveting and soldering. Acetylene burning and electric arc welding may be employed when attaching fasteners to metal support poles.

Service calls make up a large part of electric sign service. When a sign fails to work properly, a repair person is dispatched to put it back into service. This worker must diagnose the problem quickly and follow up with repairs. To do so requires a comprehensive understanding of electrical circuitry as it relates to electric signs, of transformer theory and application, and of gases and their use in lighting signs. The repair person must also understand such mechanical elements as gears, drives, bearings, and similar elements of revolving signs.

Mechanical devices and parts that may have to be replaced include defective wiring, transformers, flashers, time clocks, sockets, tubes and lamps, and even electric motors.

Electric sign workers must know the National Electrical Code, particularly the sections related to their work. They must also be aware of local building code requirements for signs and outdoor display structures. Local codes generally deal with such aspects as sign size, safety, attachment, and anchorage requirements.

Preventive maintenance is another major duty in this occupation. Many firms purchase service contracts at the time they buy an illuminated advertising display. Other companies contract for sign maintenance at a later time. In either case, periodic routine maintenance is done on such equipment.

Those with a creative flair may find electric sign work appealing for the element of customer service, for they may be called upon to suggest ways to improve the attractiveness or effectiveness of an advertising display.

The job requires an investment of about $250 for the tools of the trade, which are basically the hand tools used by an inside construction electrician. Test equipment and larger tools are usually supplied by the employer.

To succeed in this field you should feel comfortable standing and working at some distance off the ground. Workers often work from scaffolding, a motorized lift, or an aerial bucket. At times, the use of safety belts and lines may be required. Workers should be agile, with a good sense of balance and the ability to climb.

The next time you take a trip, notice the prevalence of illuminated signs surrounding cities, shopping centers, automobile dealers, and interstate highway interchanges. Your own experience will confirm the potential for job growth in the electric sign industry.

OTHER ELECTRICAL OCCUPATIONS

While smaller in number, several other occupations require electrical training and skills. For example, the

International Brotherhood of Electrical Workers (IBEW) represents communications workers employed in the radio and television broadcasting industries, the cable television industry, the recording industry, and the telephone industry. Electric railroad systems require the skills and training of operators and maintenance workers. The electrical manufacturing industry provides thousands of jobs building commercial products, home appliances, lighting fixtures, communications and data processing equipment, electric motors, medical equipment, and so on.

Electrical workers design, build, install, and service the systems, products, and services that make our lives easier. This chapter has shown you the wide choice of careers open to those with technical training in the electrical field. Perhaps you have singled out one area as particularly interesting. If not, don't worry. Because electrical theory is the foundation of each specialty, people in these careers find it relatively easy to transfer their skills from one field to another.

4

The Role of Labor Unions

Labor unions have long played an important role in protecting the rights of American workers. The International Brotherhood of Electrical Workers (IBEW), founded more than a century ago, continues to safeguard the lives and livelihood of approximately 750,000 electricians and electrical workers in the United States and Canada. The IBEW is one of the largest labor unions in the world, with divisions including construction, manufacturing, communications, and power generation.

The IBEW cites the combination of antiunion sentiment, foreign competition, and technological change as influences contributing to a general decline in union membership that began in the 1980s. That decade also saw growth in nonunion electrical contractors, while many manufacturing operations moved out of the United States, eliminating additional electrical jobs. Another factor related to union decline was the court-ordered breakup of American Telephone and Telegraph (AT&T), affecting both telephone company employees and workers in AT&T manufacturing plants.

While political and economic influences have brought about a decline in union membership since the era of the 1980s, when the IBEW boasted a million members, the union still exerts a strong influence in determining how many people enter apprenticeship programs and how they are trained. Union membership offers a measure of job security and access to pension plans, health insurance plans, and other fringe benefits. The labor union exists to negotiate with contractors (employers) on behalf of its members. Industry-wide agreements cover such important topics as pay scales and working conditions.

HOW UNIONS EVOLVED

The union movement has a long history in the United States. As early as 1778, workers began banding together in an effort to secure better wages, minimum pay rates, shorter hours, apprenticeship standards for crafts, and the advancement of union labor. Workers who struggled to form craft unions did so in the face of stiff employer opposition and government interference. In fact, unions were prosecuted as "conspiracies in restraint of trade" under English common law doctrine. Still, by 1859, the Stonecutters, Hat Finishers, Molders, Machinists, and Locomotive Engineers had all founded national organizations.

By 1850, the workday, which earlier in the century often started at sunrise and ended at sunset, was shortened to ten hours. Most skilled craftsmen were earning $2 per day or more by 1860.

The fifteen years that followed the Civil War represented an important era in the American labor movement. It was a time when fourteen new national unions were formed and the campaign for an eight-hour workday bore fruit. After years of interfering with labor unions, the government finally began to concede on several major points. In 1868, Congress adopted the eight-hour workday for federal employees.

THE IBEW IS BORN

In its infancy, the electrical industry was a hazardous place. Yet the excitement of being part of a growing and challenging field attracted many workers, despite the necessity of working high overhead for as long as twelve hours at a stretch. Public demand for electricity meant that crews were stringing line seven days a week regardless of weather. It was not uncommon for a lineworker to earn $0.15 to $0.20 per hour, or $8 for a demanding week's work.

Safety standards and supervised training were unheard of; young men learned their trade by trial and often fatal error. In some regions, half of all linemen died in job-related accidents. The mortality rate among electrical workers was double that of other industries.

The year 1890 brought an exhibition of the wonders of electricity to St. Louis, Missouri, where electrical workers from across the country gathered to wire buildings and displays. During the gathering, workers exchanged information about the dangers, long hours, and poor pay that marked their workdays. Recognizing the need for change, a move toward organization grew out of that St. Louis gathering. The first union of electrical workers was chartered as the Electrical Wiremen and Linemen's Union No. 5221 of the American Federation of Labor.

Seeking strength in numbers, the president of the local set out to organize workers of the widespread telephone and telegraph companies, power companies, electrical contractors, and manufacturers of electrical equipment. By year's end, the fledgling union had locals in more than a dozen major cities, and this at a time when public sentiment was in opposition to unions.

The union had a tough time of it in the early days. Employers were hostile, and the country was in a severe economic depression. After working so tirelessly to found the union and establish death benefits for electrical workers who died on the job, the union's first president, Henry Miller, was killed while repairing damage to the Potomac Electric Power Company's lines following an electrical storm. According to union archives, Miller received a shock, toppled from a power pole, and struck his head. After just three years of distinguished leadership, the union was without a president.

The first full-time, paid president of the union, Frank J. McNulty, took office in 1903. A decisive individual, McNulty was determined to build on the inherent sense of responsibility that electrical workers brought to their work. This sense of responsibility has been a focal point guiding all IBEW negotiations, gaining the Brotherhood a reputation among employers for not violating agreements.

In 1920, the Council on Industrial Relations for the Electrical Contracting Industry (CIR) was founded, represented equally by six delegates of the National Electrical Contractors Association and six members of the IBEW. This council is charged with settling major differences between labor and management in the industry. It functions so well that only one CIR decision has been violated in sixty-five years, earning the IBEW the title of "strikeless union."

The early 1900s marked a period of growth for the

Brotherhood. Canadian electrical workers joined the ranks in 1902, prompting members to adopt the name International Brotherhood of Electrical Workers, which has served to this day.

The IBEW currently represents nearly one million workers across the United States and Canada, along with members in Puerto Rico and the Panama Canal Zone. Its members are employed in every part of the industry. A union publication notes some of the jobs IBEW members perform. The list includes:

- Supplying energy at electric generating stations in the United States and Canada
- Inside and outside construction
- National defense for both nations
- Worldwide communications
- Railroads and urban mass transit systems
- Electrical manufacture
- Space research
- Healthcare

Industry branches of the Brotherhood include outside construction and utility workers, inside electrical workers, communications workers, railroad and Pullman electrical workers, electrical manufacturing workers, government workers, and others in varied occupations associated with electric power.

Union negotiations with the electrical industry have significantly raised the standard of living for workers. The length of the workday and the workweek have been scaled back; most agreements provide for health and welfare plans for workers and their families; and vacations and retirement benefits are now the norm, as are paid holidays. Safety remains a primary concern, and many of the hazards of the industry have been eliminated. IBEW members are well trained and adequately paid during the process.

The IBEW is also dedicated to leadership in promoting both economic and social justice for all citizens. One million members strong, the union has adopted as its motto, "A better way of life through the IBEW."

Dues

In return for union representation and services, members pay a one-time initiation fee and monthly dues. Dues cover the cost of running the organization, including union representatives, attorneys, research workers, and an administrative staff. Each union has officers elected by the membership and draws up bylaws reflecting grassroots issues. Members run the union hall, deal with management on a daily basis, locate work for members, and assign them to jobs.

To be eligible for membership benefits, a worker must keep dues paid up. Union members carry a dues book or card to confirm that their membership is in good standing.

As said earlier, most of today's union workers "graduated" to membership after completing a union-sponsored apprenticeship of three or four years. Doing so requires passing a test administered by the union local and paying an initiation fee, which may amount to several hundred dollars.

Craftworkers who have attained journeyworker status by some method other than apprenticeship may also apply for union membership by taking the admission skills test and paying the initiation fee.

Because unions constantly work toward job security for their members, one of their goals is to maintain a balance between the number of available jobs and the pool of available workers. The number of apprenticeship slots made available reflects this supply-and-demand issue.

To a certain degree, union electrical workers, particularly those involved in the construction trades, are shielded from seasonal layoff through membership. If the downtime is lengthy, the union local helps to find a new assignment, and a number of unions provide SUB pay (supplemental unemployment benefits) during a furlough. The network of union locals also serves as an information exchange about job opportunities in other areas, should a member be interested in relocating.

Retirement Options

The IBEW was among the leading trade unions to acknowledge the need for pension plans. Financed by a contribution of $0.37 per month from members, the union plan was established in 1927. As initially drawn up, the plan provided for retirement at age sixty-five after twenty years of membership. The pension was $40 per month for life, a sum that provided a life of dignity and comfort for retirees of that era.

Over the years, benefits have been improved to reflect changing economic times. Today the IBEW Pension Benefit Fund offers a spouse's option, disability benefits, and death benefits.

In 1946, an agreement between the IBEW and the National Electrical Contractors Association (NECA) established the National Electrical Benefit Fund (NEBF), representing an effort to have union electricians and their employers share part of the expense of paying pensions to retired IBEW members. Contractors began contributing 1 percent of total payroll to a national trust fund, from which pension payments were drawn. In 1957, however, the National Labor Relations Board ruled that it was illegal to limit pension benefits to IBEW members. Since then, two separate funds, the NEBF and the IBEW Pension Benefit Fund, have carried out the goal of providing pensions for retired electrical workers.

The IBEW maintains a Department of Special Services, dedicated to identifying more efficient ways to serve the needs of members who have retired or are approaching retirement. One important function is sponsoring informational retirement planning programs.

Independent electricians, that is, those who are not union members, may or may not be covered by pension plans. Those who are employed by private industry may have certain fringe benefits, including a pension, as part of the overall conditions of employment. Others make their own retirement savings plans, usually by investing a portion of their earnings in tax-deferred savings known as an Individual Retirement Account (IRA).

Fringe Benefits

Union electricians are represented by their bargaining units in the matter of securing fringe benefits. The union local bargains for such items as paid sick leave, vacation time, holidays, and working conditions. Thus, such matters vary from local to local across geographic regions. To find out more about this aspect of an electrician's career, contact the union local nearest you.

Electricians and electrical workers employed in private industry generally receive a benefit package similar to that provided for other workers in the company's employ. Typically, health insurance, paid vacation and holidays, disability income protection, and retirement benefits are covered in such packages.

Although this chapter has focused on the IBEW, a number of maintenance electricians belong to the International Union of Electronic, Electrical, Salaried, Machine, and Furniture Workers. Most line installers and cable splicers who do not belong to the IBEW are represented by the Communications Workers of America.

5

Broadening Horizons for Women and Minorities

Despite the political controversy over affirmative action programs, construction trades are hungry for new skilled workers. The decade ahead promises many opportunities for women and minority workers who are trained as electricians or electrical workers.

In a study of critical issues facing the homebuilding industry, the National Association of Home Builders discovered that the construction industry will need to recruit more than 340,000 workers. Now, and well into the twenty-first century, almost half of the new entrants in the labor force will be women. Minorities will account for 33 percent of new workers—double their current share of the workforce.

Home builders are actively recruiting women and minority workers with positive attitudes and the willingness to learn. And these figures represent just the homebuilding industry. Commercial builders and the electric power industry will face the same competition for skilled workers. In other words, there has never been a better time to enter this career field.

Because electricians command good pay and the respect of other workers, it is no surprise that many women and minority candidates seek this career. Electrician apprenticeships were awarded to 1,646 women and 4,951 minority candidates in 1990. There were about 200 more electrician apprentices than carpentry apprentices, the second most sought-after craft training.

Remember that electrician apprenticeship also trains people to work in electrical maintenance, marine, and power plant occupations. These industries are not as likely to compile statistics on recruitment programs, but they are subject to the same labor market forces as the construction industry.

The combined weight of several factors will keep the employment gates always open and widening. Government supervision of industry hiring practices is a major influence. Contractors who bid on government contracts are already working with target figures in hiring. Enlightened attitudes on the part of unions and employers are taking effect, and the exploding of myths about gender and the nature of work are having a significant impact. As these trades compete with other industries for the best and brightest among a new generation of workers, salary increases and more attractive working conditions will result.

Equal opportunity is a cornerstone of government-funded training programs such as the Job Corps. Both federal and state agencies are working closely with employers, labor unions, and vocational schools to improve the quality and availability of apprenticeship training. Many adult literacy programs also offer job-training programs with an emphasis on nontraditional careers.

The federal Job Corps program has trained more than two million young people, including women and minorities, for nontraditional programs. Electrical wiring is one of eleven building trades taught at Job Corps centers across the country.

The Home Builders Institute, the educational and training division of the National Association of Home Builders, recruits both women and minorities for its Craft Skills Program, which is operated by more than 100 builder associations and which serves as a gateway to the trades for nearly 50,000 young men and women. Electrical wiring is one of its most sought-after programs.

There are also more than 100 community-based employment training programs dedicated to helping women get better jobs. Most of these groups offer counseling, ability and aptitude testing, physical conditioning, assertiveness training, hands-on skill training, and job search services. Such programs generally last from fourteen to twenty weeks and are open to persons age eighteen and older.

To better appreciate what such programs have to offer, let's take a closer look at a few. Wider Opportunities for Women, Inc., popularly known by its initials—WOW—is based in Washington, DC. Its program includes interview skills, assistance in filing applications, assessment of one's skills, and exploration of career options. WOW also operates a program that prepares economically disadvantaged women for construction jobs.

The national network of YWCAs has a long tradition of assisting women in seeking more challenging, better-paying employment. More than a century ago, YWCAs took on the task of teaching women to type at a time when such a job was considered "too demanding" for them. Current efforts focus on identifying job opportunities, recruiting women workers, and providing necessary support services. You can find the nearest YWCA in the phone book, where you will also find the numbers for useful government programs with job-search and job-training services. If you are a displaced homemaker or eligible for adult education or literacy programs, help in improving your employment prospects is probably also available to you.

Women who own or manage an electrical contracting firm are represented among the more than 8,500 members of the National Association of Women in Construction, with headquarters in Fort Worth, Texas. This organization provides job referrals, takes part in career recruitment, shares industry knowledge and ideas, and offers construction education, including scholarships for women preparing to enter the industry. To contact NAWIC, call (800) 552-3506 or write to the NAWIC Executive Office, 437 South Adams Street, Fort Worth, Texas, 76104.

Another sponsor of job-training programs is the National Urban League's Labor Education Advancement Program (LEAP), which you may contact at 500 East 62nd Street, New York, NY 10021.

The Women's Bureau of the U.S. Department of Labor has recently compiled a directory of 125 nontraditional training and employment programs serving women. It can be purchased from the U.S. Government Printing Office, Superintendent of Documents, Mail Stop: SSOP, Washington, DC, 20402-9328.

To learn more about recruitment programs for minorities and women, contact the Job Service office listed in your local telephone directory.

A Woman Electrician

A career change took Anne away from the remodeling business and into the world of the journeyworker electrician. The choice proved to be satisfying in many ways.

"I was working with an older fellow who remodeled kitchens and baths, and one of my jobs was to hire the subcontractors—the plumbers and electricians. One of the electrical subcontractors is now my husband, the owner of an electrical contracting business, who helped to convince me that I should enter an apprenticeship. The man I had been working for was planning to retire, so I would have faced some changes anyway. After considering all the pros and cons, I decided it was definitely worth a try. It took me three years to fulfill the 8,000 hours required of an apprentice. I worked by day and took classes by night, 300 hours of electrical theory. Part of that time I was also working as a waitress. When I took the Massachusetts journeyworker electrician exam, I was the only woman in a class of 100 apprentices. There was one other woman in the theory class, but she was working in a factory at the time and may not have needed to be licensed for the work she was to do."

Much of the exam concentrated on electrical theory and how it is applied. Passing the test meant that Anne could apply for a journeyworker electrician's license. She pays $30 every three years to keep her license current. Her husband must maintain both the journeyworker's and master's licenses to participate in the day-to-day work of his contracting business. Earning the master's license requires passing yet another exam. The combined licensing fee amounts to $75 for three years in Massachusetts.

Why be an electrician? "It's a very respectable way to earn a good living, and there are great possibilities for advancement. There will always be work; that's never been a worry for me. People spend thousands of dollars and several years to prepare for some professions, only to find that there are no openings when they graduate."

Anne is an independent electrician, which means she is not a member of a labor union. The family contracting business primarily handles residential wiring, but it also has clients from the commercial, industrial, and manufacturing sectors, along with several restaurant clients.

"It's not always clean work," Anne points out. "Most wires run from the basement to the attic. In an old structure, there can be years of accumulated grime. Existing wiring is often covered with soot. Even when you are upgrading the electrical system for a home, it's dirty work. My regular work clothes are jeans and sturdy shoes. I wear a hard hat when it's required.

"My idea of the ideal wiring job? Definitely a new home. I often get to design the wiring. It's interesting to match the wiring to the lifestyle of the family building the house. Planning where to place the outlets and switches is different for different styles of living. For example, take the master bedroom. Most people have night tables of some type beside or between the beds and a lamp on each table. Thoughtful design will wire one or more lamps to the light switch by the door. That way the lamp can be turned on from the doorway, without having to cross the room in the dark or turn on an overhead light.

"Much of the same sort of planning can make a home security system or entertainment system function more conveniently. Security lights and an alarm controlled by a dual system might have controls in the foyer or near the kitchen, but switches in the master bedroom are really convenient. Chances are, if you want to turn the security lights on, it will be after dark. The panel can be wired so that you don't have to get out of bed or leave the room to reach it."

In addition to the challenge and satisfaction of working on an original wiring design, the electrician who works on a new home is drilling into new lumber, Anne points out, "not dirty old lumber with the potential for giving you an infection from the splinters you get while working." Anne has grown accustomed to getting dirty on her job. "Grimy, that's the only word for it, but it has its rewards, too."

"Being an electrician is certainly not boring. You're doing something different every day," she says. In addition to assignments that have taken her to homes, businesses, restaurants, stores, and factories, Anne has also done electrical repairs and installations on farms surrounding her Massachusetts hometown.

"Strict codes apply to electrical work around dairy barns, since any stray current adversely affects the cows. They are very susceptible. Besides, the high humidity in that setting creates a corrosive atmosphere. You have to meet the minimum standards of the National Electrical Code, and most states have enacted their own, usually more stringent, codes. Most communities have a wiring inspector who can supersede the national code by whatever he chooses to mandate."

Anne addresses the work habits of a quality-minded electrician. "Neatness is a big thing in this field. Sure, it's a lot faster to drape wire and cable around things, especially when someone else is paying for the materials, but that sort of sloppy job is not very professional. There's something almost elegant about a tidy junction box. The best electricians take pride in their work and want to do a neat job with no flying splices."

Without a doubt, workplace safety is the top priority with this electrician. "Often I get called to a job where other people have fiddled with the wiring in the past. Some of the jobs have been handled well, but others make you shudder. Most of the time, I can avoid working around live lines, but I am particularly careful whenever I spot signs of amateurish wiring. There's the distinct possibility that someone has created an electrical monster that's lurking within the walls, just waiting for you. I have more fear, or perhaps you should call it a 'healthy respect,' for the power of electric current than I did before I became a professional electrician. My rule is, 'Don't touch anything' until after a careful inspection. I'm always thinking, 'Will I get a jolt from this one?' These situations are particularly evident in residential wiring. You often see horrendous things done, really dangerous wiring that is a fire hazard. You never know what you could run into, and I'm not anxious to suffer the consequences of someone else's carelessness or inexperience."

Anne ruefully admits having been badly shocked. "And it happened in my own kitchen." Anne explains that she was preparing dinner one evening, using an electric skillet and a metal spoon to stir the contents. "I guess there was a short in the frying pan, and my free hand accidentally came in contact with the edge of the stainless steel sink nearby. I took 120 volts directly across my heart. It knocked me down and I went rigid, but I was lucky. It's ironic that the worst shock I ever received happened off-duty, in my own kitchen. Oh, I've had little 'buzzes' or electrical bites while working, but nothing like the experience with that electric frying pan."

Anne says she is slightly claustrophobic, which has caused her some anxiety on occasion. "There was the time I found myself squeezed into the narrow space inside a wall, pulling wire," she recalls. "I had to face a little sense of panic, but I got through it. On another job, I got my arm stuck in a wall while fishing wire. That bothered me, too. Luckily, I have no fear of heights. That's something that would definitely affect the career of an electrician. Also, it's not always a simple matter to place ladders securely. I remember the time we were wiring the safety controls for a freight elevator. The controls were necessary in case a load were to shift or the elevator door to jam. At any rate, we were running a number of conductors in a very, very tight space. We had the ladders nailed to the sides of the shaft, and the space for fishing the wires was so tight that our only means of getting them through was to grease them—we used cooking shortening, the kind you'd put in a cookie recipe. That was one of the most demanding assignments I've been on."

Although Anne and her husband are self-employed, they are conditioned to regular business hours of 7:30 AM to 5 PM, and they try to avoid working weekends. However, the nature of their business means there will be emergency service calls " at the height of a storm or in the dark of night. It's not unusual to receive a phone call at 2 AM," she explains.

"Those are times when extreme care is called for. A couple of years ago, an earthquake struck this area, affecting some underground electric lines. We were trying to locate lines and boxes, to make repairs. We advised the property owner to have a new service installed, but he was reluctant to make that investment, so we did as instructed and just made repairs. A week later during a torrential rain, the system failed again. It was a dangerous situation, and I worked with a crew from the utility company to install a temporary aerial line until conditions improved and we could replace the underground service."

Soon after completing her apprenticeship, Anne got a chance to prove both her strength and resourcefulness. Her husband, a road-racing enthusiast, had left for a competition early one Saturday morning, taking the company truck with the ladders and much of their equipment. That morning Anne got an emergency call from an important client—a storm had taken down the entire electric service at his home.

"I discovered he had a painter's ladder, so I drove out to have a look at the job. There was an electrical supply company with Saturday morning hours, so I got the materials I needed. It took a lot of effort, carrying that wire up the ladder and getting everything back into working condition, but when my husband got home he was delighted at the initiative I'd shown. Coming through in an emergency is important to a small company. It saves clients, who are always tremendously grateful that you helped them when they really needed it.

"Any time you work on the electrical system of an occupied structure, it requires two visits. First you disconnect and make repairs or installations; then you schedule the electrical inspector's visit. After the job passes inspection, you go back to reconnect the system."

Does being an electrician require any special physical training or strength? Anne concedes that conduit bending requires a certain degree of upper-body strength, but nothing that should deter an able-bodied person, male or female. "Carrying a reel of wire is strenuous, I'll admit, but I've learned some of the tricks of working alone, such as to roll the reel instead of trying to carry it."

One aspect of the work is a challenge to all electricians, male or female. "That's the weather," Anne explains. "During the winter months, you try really hard not to schedule outside work. There's no way you can do the job in gloves, and your fingers and hands get numb from the cold. They're stiff, and it's hard to work well under those conditions. Your hands get awfully rough under such conditions, too, but that's one of the tradeoffs of a well-paid job."

As a successful nontraditional worker, Anne has been called upon as a guest speaker at vocational training schools. "I once addressed a group of female high school students who were enrolled in a trade school. They had a lot of questions about how I was treated by fellow workers at job sites. Apparently they had some schoolmates who were not very tolerant of female coworkers. I was able to reassure them that professionalism on their part, and a degree of maturity they could expect from their male coworkers, would resolve the issue. I can't see any insurmountable barriers to a woman's becoming an electrician."

In Massachusetts and surrounding states, journeyworker electricians are required to take regular renewal courses to keep up their license. "The first time I went through this process", Anne recalls, "I got the distinct impression that the instructor had singled me out for a kind of oral test, to determine whether I actually worked as an electrician. Of all the people in the room, I was chosen to answer about 80 percent of the instructor's questions." Aside from this grilling, Anne has never been singled out for harassment as a woman in her blue-collar profession.

"Frankly, I think women are particularly well suited to work as electricians," she comments. "They may tend to be a little more cautious in their approach to electric current than men. At least, that's been my experience." Regardless of gender, she thinks that good tools, good training, caution, and experience are the elements essential to a long and rewarding career.

Like all independent, self-employed workers, Anne and her husband must purchase their own health insurance coverage, pay a higher rate of Social Security taxes, and meet the cost of fringe benefits a worker receives from an employer. For this reason, Anne has occasionally thought about seeking work elsewhere. "I've seen the maintenance electricians employed by a college in our area. They drive around campus in a pickup truck and don't even wear a tool belt. Maybe that's the sort of job I'll get when I'm older and not looking forward to the physical demands of contracting."

Another option worth exploring is to move into a supervisory position, although to do so would require additional training. "There's no denying that this can be a physically demanding way to earn a living," she says.

Anne knows of at least one female master electrician at work in the Boston area. To move ahead to that level in her home state, Anne would need only to sit for the master electrician's exam, since she has already fulfilled the minimum of one year's experience as a journeyworker.

6

Apprenticeships: Learning by Doing

Let's take a closer look at an electrician's training. A formal four- or five-year apprenticeship is considered the best way to become an electrician, so a significant number of today's electricians are pursuing this path, which is a combination of classroom instruction and on-the-job training.

The length of apprenticeship varies significantly from one trade to another, but programs range from one to six years, with three to four years being typical.

As you might imagine, there is considerable competition for free job training that pays a wage while you advance toward a well-paid career. You and the other candidates for apprenticeship slots may have to take a test and participate in a personal interview as part of the selection process. Most programs also require that you pass a physical exam.

Most apprenticeship applicants take a test called the Specific Aptitude Test Battery, which has nine parts. Typically, two test areas are administered, chosen from the following list: general learning ability, verbal aptitude, numerical aptitude, spatial aptitude, form perception (the ability to perceive small details), clerical perception (the ability to distinguish important details), motor coordination, finger dexterity, and manual dexterity.

The aptitude test helps define a candidate's specific skills and abilities. This has proved to be an effective method of matching people with job training at which they will succeed, since they are likely to enjoy the work and do well.

Employers and unions put up the money to operate apprenticeship programs, and employers devote the time and talents of skilled workers to train apprentices. Therefore, both want to recruit the best and the brightest apprentices.

The training programs are also subject to Department of Labor rules concerning equal opportunity. Women, blacks, Hispanics, and other minorities must get a fair share of apprenticeship slots. In addition to the electrician's apprenticeship, many electric utility companies offer an in-plant apprenticeship for line and cable installers. For more about these opportunities, see chapter seven.

Some trade school graduates have an advantage in applying for electrician apprenticeships, since they are already familiar with shop safety, materials, and tools. They may be granted higher starting wages and face a shorter training period.

Apprentices who are veterans of U.S. military service have another advantage as they pursue their job training. The Veterans Administration provides a monthly training stipend, or payment, for on-the-job training that meets certain requirements. These payments gradually decline, offset by rising wages.

PERSONAL QUALIFICATIONS

The electrical trades require a unique combination of talents and natural aptitudes. You should be skilled in the use of tools and have the ability to master theory.

Of even greater importance, the trainee should have a strong interest in mastering the technical information, and a desire to achieve.

Apprenticeship applicants should be physically fit and able to perform manual functions in a safe, productive manner. Although much work is done indoors, candidates must also be alert, agile, and physically capable of working outdoors under adverse weather conditions. They should be able to drive a lightweight truck or van and be familiar with the locality in order to find work sites quickly.

A solid foundation in reading, writing, and math is required of all apprentices. Candidates who have done well in such high school courses as shop, math, drafting, and physics also have an edge. If you are still in high school, you would be wise to take as many science and math courses as possible. If vocational-technical training is offered, courses in blueprint reading, electronics, and basic wiring will also be helpful. Good study habits and a good attendance record will help you earn a slot in an apprenticeship program.

Apprenticeship candidates are evaluated on the basis of interest shown, attitude (toward hard work, authority, and teamwork), and such personal traits as appearance, assertiveness, sincerity, dependability, character, and personal habits.

Points are assigned in each ratings category, totaling 100. Scores for interest, attitude, and personal traits come from the interview. The total point scale is entered on "the register"—a waiting list for apprenticeship openings—and you will be notified of the scores given.

ASSIGNMENT TO PROGRAM

The next step is to wait for an opening. Job experts note that the wait can be months or even years, depending on the number of openings and the number of applicants with ratings higher than your own. Estimates indicate that there are eight applicants for every construction apprenticeship slot.

According to the National Joint Apprenticeship Training partnership, there are approximately 30,000 apprentices learning the electrician's trade at any given time in the United States. With more than 300 training centers offering a combination of classroom and on-the-job training, young men and women are learning skills to last a lifetime.

The state of the economy and geographic location also have an impact on apprenticeship activity. As you might imagine, when skilled workers are being laid off, the chances for new trainees to get a start decline sharply.

If you are not called within two years, you have to reactivate your file by reapplying. That does not mean you have to go through the interview again, but chances are you'll improve your rating score if you do.

THE APPRENTICESHIP AGREEMENT

Before on-the-job training or classroom work begins, a formal apprenticeship agreement is drafted and signed by the trainee and the training organization. Apprentices who are under twenty-one years of age must have a parent or guardian sign the agreement. This document legally assures the apprentice of training and preparation for a career so long as the apprentice lives up to the requirements stated. You receive an ID card to carry, and it is updated regularly to reflect your progress toward journeyworker status and wages. On payday, the contractor who employs you issues a check for your wages. An effort is made to provide fringe benefits to apprentices on the same basis as to journeyworkers.

Unfortunately, not every apprentice completes training. If some circumstance forces you to drop out of the program, you can resume training later with credit for the work already accomplished. Apprentices should be aware that a pattern of unsatisfactory performance is grounds for dismissal.

APPRENTICESHIP SUBJECTS

Apprentices are required to master a wide variety of formulas and mathematical computations. Candidates are evaluated on the basis of academic performance and potential, health status, attitude, and enthusiasm for the work. During the course of their training, apprentices usually work for several contractors, becoming familiar with a variety of electrical jobs.

For example, the construction electrician's apprentice program covers four years (8,000 hours) and is a combination of supervised on-the-job training and classroom instruction, the latter covering 144 hours per year. Many programs offer some study of electronics. On the job, you are guided by an experienced electrician. At first the tasks are routine, such as learning to drill holes, setting up conduit, and setting anchors, while you learn to handle the tools of the trade. You progress to measuring, bending, and installing conduit and connecting and testing complete wiring systems. Here is a brief summary of apprenticeship training:

I. Residential (2,000 hours). Single-phase service and metering, remodeling, installation of equipment and appliances, light fixtures, receptacles and switches, security systems.

II. Commercial-industrial (5,000 hours). Electrical services, 240 and 480 volt; metering, polyphase and current transformers; installation of conduits and outlets, wiring beneath concrete slabs and masonry, steel construction, buss-duct systems, under-floor ducts, and via metal raceways; installations that are vapor-proof and explosion-proof; circuiting for light and power; various types of motors and their controls; transformers, applications, and connections.

III. Specialized work (1,000 hours). Welding, management-employee relations, customer-employee relations, electronics systems, communications systems, and fire alarm systems.

Total: 8,000 hours.

The following is a breakdown of the training schedule for the various apprentice specialties.

Industrial Maintenance

I. Electrical construction (2,076 hours). Begins with safety instructions, moving on to running wires and making hookups; installing lighting, power circuits, and power and control equipment; installing motors and generators; laying out the job from blueprints; and selecting materials.

II. General maintenance (1,766 hours). Beginning with safety instructions, the apprentice learns to diagnose trouble in lighting and power circuits, to check and repair electrical equipment, and to adjust and repair welders.

III. Cranes and elevators (520 hours). Includes safety instructions to check, repair, and adjust limit switches and safety devices, locating and repairing faulty electrical equipment.

IV. Electrical repair (1,650 hours). Emphasizes safety instructions, AC and DC motor repair, repairing controllers and heating appliances, building and repairing transformers and welders, and making drill repairs.

V. Powerhouse instruction, including substation construction (850 hours). Focuses on safety instruction, plus heavy cable installation and layout and installation of heavy conduit and ductwork; installing master distribution cabinets; cable splicing; wiring busways and bus, switchboards and switchgears, transformers, motors, and generators; connecting instrument transformers, meters and relays; and checking and testing circuits.

VI. Electrical maintenance of powerhouse and substation (466 hours). Covers safety instructions for servicing main and auxiliary circuits, servicing equipment, adjusting timing of relays, testing and reconditioning the transformer and oil switches, testing coolants for transformer and switches, and testing of all circuits.

VII. Related instruction, 672 hours.

Total: 8,000 hours.

Electric Motor Shop Journeyworker

I. Shop work (8,000 hours). Covers shop errands and orientation, cleaning electrical machinery and auxiliary equipment of various kinds, stripping old windings, cleaning bar coils and tinning leads, appliance and fan repairs, locating and repairing trouble in fractional H.P. motors (AC); winding field coils, DC armatures, transformers, coils, and solenoid coils.

II. Also, testing field coils and armatures for shorts, opens, and grounds; troubleshooting on various types of equipment in the shop; winding and forming armature and coils, with redesign of same; commutators and general rebuilding of same; batteries and battery-charging equipment; starter and control repairs; winding and repair of armatures, rotors, and stators; repairs of brushes, brush holders, and motor leads.

III. General machine shop practices; assisting with troubleshooting in field; switchboard work; welding, spot and flame; taping and insulating coils, also bar windings; testing various types of electrical apparatus. Total: 8,000 hours.

Marine Electrician

I. Wireways (1,000 hours). Layout of raceways, raceway prints; making and installing hanger tubes; cutting threads, bending and installing conduit; installing cables; stripping cables.

II. Power wiring (2,000 hours). Print reading, installing motors and controllers, connecting motors, controller switches, and circuit breakers.

III. Marine lighting (500 hours). Layout print reading, installing fillings and fixtures, making joints, soldering, taping, and splicing.

IV. Interior and other communications (1,000 hours). Connect bells, batteries, alarm connections, and gyros; telephones, engines, order telegraphs, radios, and follow-up systems.

V. Installation of generators and switchboards (500 hours).

VI. Yard maintenance (2,000 hours). Lights, transformers, and connections; AC power, motors, cranes.

VII. Opening electrical equipment (1,000 hours). Testing of circuit continuity, voltage, amperes, measurements.

Total: 8,000 hours.

Electrical Lineworker (Light and Power)

I. Prerequisite experience: Loading and unloading tools and supplies on trucks; setting poles and anchors; sending tools and supplies to lineworker on pole; other assigned ground work. Names and uses of tools and materials; care of tools and rubber goods; how to tie rope knots and slings; how to make various partial assemblies of materials on the ground before installation on the pole.

II. Line construction: Setting, guying, and anchoring poles and pole-top work; use and care of climbing tools; pole climbing skill; safety precautions, including use of rubber gloves and blankets; first aid and safety methods; teamwork, sequence of operations; public relations and safety; how to arrange for short service interruptions; keeping job and work order accounts. Also, pole line construction methods: span lengths and sagging-in, hardware; tree clearance; reading electric drawings and sketches, and ability to make a legible sketch or diagram.

III. Building transformer structures, installing transformers and protective equipment, and connecting to lines; construction methods for transformer structures; how to connect and phase out transformers and lines; knowledge of connections commonly used, and how to work on energized lines up to 4,500 volts.

IV. Streetlight installation. Series and multiple circuits; methods of making installations.

Term: Four years.

Powerhouse Electrician (Light and Power)

I. Orientation: Learn physical location of electrical equipment; learn power plant terminology and company organization; become familiar with symbols and switching diagrams.

II. Maintenance of electrical equipment, including oil circuit breakers, lighting equipment, and starters; knowledge of routine maintenance such as oil changes, cleaning, and inspection; knowledge of removing electrical equipment from service and obtaining clearances before switching; proper care of storage batteries; knowledge of simple lighting circuits and methods of repairing lighting fixtures.

III. Check and maintain safety equipment; familiarity with various types of fire extinguishers and methods of refilling; first aid and safety methods; proper use of rubber gloves and blankets; knowledge of all types of fuses and fuse cartridges in use; understanding of methods of fighting electrical fires.

IV. Perform shop and plant duties, as required—care and use of hand tools; knowledge of soldering, brazing, and elementary welding (gas and electric types); care and use of power tools (lathe, drill press); knowledge and use of insulation materials, and knowledge of checking insulation by use of megger; use and interpretation of a "growler."

V. Maintain records, as required—knowledge of

purpose of records; basic accounting for proper distribution of materials.

VI. Install conduit and wiring for any electrical equipment—method of installing conduit, knowledge and proper use of conduits; familiarity with the use of proper cable and wire as related to size and insulation; method of handling wire in conduit; knowledge of conduit sizes and permissible circuits in any size.

VII. Install and connect transformer—methods of installation and connection; knowledge of connections and circuits of current, potential, and power transformers; theory and safety precautions to be observed; phasing between banks; checking voltages of various connections.

VIII. Installation and maintenance of regulators—theory of induction and voltage regulators; method of control; routine care, methods of installation, and maintenance.

IX. Installation and maintenance of rotating machinery—theory, care, and installation methods of AC, DC generators and motors (includes a single-phase, polyphase, synchronous, and induction constant and variable speed for AC and series, shunt, and compound wound DC machines).

X. Install meters, relays, and controls—basic theory of watt-hour meter, voltmeter, ammeter, and wattmeter; analysis of control diagrams; knowledge of relays and remote controls; adjustment of elevator controls; knowledge of proper uses of voltmeters and ammeters; understanding of interlocking controls; knowledge of operation of industrial instruments such as temperature recorders and remote flow meters; knowledge of location of grounds and short circuits and ability to apply emergency correction; understanding of when to use "stay put" or other types of push buttons, as well as voltage releases for continuity of service.

XI. Perform general plant construction—working knowledge of plant power distribution; knowledge of wiring diagrams, blueprint reading; elementary knowledge of electronics and its industrial use; methods of laying out general plant construction work; how to plan and choose switch capacities, circuit capacities, controls, and plant installations.
Total: 8,000 hours.

Electric Sign Installation and Service

I. Shop wiring—primary and secondary insulation; flashers, transformers, time clocks, switches, and fuse blocks.

II. Shop assembly—installing tube supports, bushings, housings, and mounting tubing.

III. Shop erecting—prefabrication, structural steel; roof and marquee signs; operations required; burning, punching, welding, bolting, coping, and loading.

IV. Job erecting—layout, preparation of site for installation; erection of steel structure when necessary; installation of rigging; landing and securing signs; connections to power supply.

V. Service or maintenance—troubleshooting, removing defective parts; replacing defective parts; repairing, patching, and refinishing. Term: four years.

AN APPRENTICE CLASSROOM

It is a Friday morning in September on the campus of a community college where twenty-seven apprentice electricians are meeting for their weekly seven-hour classroom session. The instructor, a journeyman electrician, began "stringing wire" about the time most of the apprentices were born. These meetings began in mid-June and will continue until each participant has logged the necessary 300 hours of classroom instruction.

Each Monday through Thursday, the apprentices report to work assignments with sponsoring electrical contractors. There they receive on-the-job training in the use of tools and materials necessary to the trade they are learning. Fridays are different. They must hit the books, tackling lessons that become increasingly challenging as the weeks go by.

"We teach electrical theory, the basic principles and measurements of electricity, AC and DC current, and basic electrical devices and circuits," the instructor, Gene, explains. "Beyond that, we study blueprint reading, job planning and material estimates, the types and sizes of wiring used for various applications, the electrical service entrance and circuits, and later, fundamentals of the electrical code."

Today's schedule calls for demonstration tests in some basic skills. While the class works at an assignment written on the chalkboard, a few students are called to one side of the room, the lab, where Gene presides over a workbench. Each student draws a slip of paper from a small file box and reads the task he has chosen at random. The first student selects a cordless, variable speed drill, a length of conduit, and several anchors. Moving to the workbench, he proceeds to measure, bend pipe, and drill holes to set the anchors, following instructions for the task he has chosen. The apprentice's movements are sure and confident, because he has practiced this routine many times at his work assignment. When the routine has been completed, the instructor signs a checklist documenting that the student has mastered each step of the process.

The second student draws the assignment of wiring a ground fault circuit interrupter (GFCI), a safety device typically installed in kitchens and bathrooms.

Next comes a demonstration of the planning required to upgrade wiring in a fifty-year-old home.

Another student must explain the steps in troubleshooting a portable dishwasher, finding a short circuit that could cause a serious electric shock.

Still another student demonstrates the use of various metering and test equipment; checks construction drawings for errors and oversights in the installation of a buss-duct system; and details the installation of a home security alarm.

"Every one of these students works hard the rest of the week using what we have covered in our Friday session. The program is carefully coordinated to make sure they have time to absorb the theory before they practice it on the job. Since the first 2,000 hours of their apprenticeship are devoted to residential electrical systems, each student works with a contractor in that field. The apprenticeship requires a total of 8,000 hours, the greater part of which is in Phase II, with its commercial and industrial applications. When these people reach Phase II, it will probably mean a change of employer. They'll work with crews in factory and commercial settings, probably helping to wire elevator shafts, heavy-duty industrial equipment, restaurant kitchens, public buildings, even traffic lights.

"For several of our students, the start of apprenticeship classes was really rough. The first few classes were devoted to testing abilities. Some needed remedial work in math, particularly shop math, and others had never before handled the tools of the electrician's trade. We had them learning to climb a ladder while wearing a tool belt and carrying a reel of wire—quite an experience for some of them. Those who were deficient in certain skills were given self-paced workbooks to take home and lengthy assignments to do between classroom sessions. I must say, their attitudes are great; no excuses, no dragging of feet. Everyone here has seen the future, and it says 'electrician' in great big capital letters. That's what they want, a secure and satisfying career."

One of the first tasks was to give the apprentices an understanding of how electricity works. "That can be pretty tricky, since the standard definition of electricity is an invisible energy capable of moving 186,000 miles per second. Although humans have used this mysterious power for more than 100 years, there is still no exact definition for it," Gene explains.

"Our first phase of study is residential electricity. Even though water and electricity should never mix, the residential systems for each have some things in common, so it's an apt analogy. Both water and electricity enter a house through a main line, which then branches off to points of use. Each system includes interrupters in the layout, so that the system can be shut down entirely or at specific, isolated points.

"Manual shutoffs are typically provided for the water line. The electrical systems can also be operated manually, but for safety, they have automatic functions that shut down in case of a dangerous overload. Valves do the job on a water system whereas circuit breakers serve the same function in an electrical system.

"Another similarity has to do with the diameter of the pipe or wire involved. The larger the water pipe, the more volume it can carry. Likewise, the larger the wire, the greater the flow of electric current. Whereas water flow is gauged at so many gallons per minute, and water pressure at pounds per square inch, the flow of electrical current is stated as amperage, while the pressure is voltage.

"Like the plumbing system in a house, the electrical layout is not really that difficult to plan or carry out, but the key to both is expertise. Do careless work or use inadequate materials, and you'll have big problems. Proceeding without due caution for your personal safety, particularly with electricity, can be deadly. If you forget to shut down the water system before you work on it, you may flood a room or get soaked. Forget to cut off the electric current and you can be killed. There's one rule that can never be overlooked—kill the current while you're working, not yourself.

"It's important for apprentice electricians to know that if connections have been made at the entry panel, the system will still be live to that point even if they have disengaged the main house switch. I also emphasize that black wires are always connected to black wires; white wires to white wires. And white wires are continuous; they are never broken for the connection except in particular circuits, and then the white wire is specially marked to indicate its unusual use."

During the mid-morning coffee break, discussion turns to the National Electrical Code, a book containing rules, specifications, and installation regulations compiled by the National Fire Protection Association. Local wiring inspectors consider it something like a bible in their work. The contents, revised periodically to keep pace with technology, is the result of experience, knowledge, and testing.

"We work only with materials that bear the U.L. identification stamp, for Underwriters Laboratories. That means they have been tested for quality and safety. I certainly hope this is a habit that rubs off on our apprentices. Do your best work, using the best materials, and you've got something that will last for a lifetime.

"Another important safety concept is that of grounding an electrical system at the entrance and continuously through the installation. This is absolutely vital. Grounding is connection of the parts of a wiring installation to a ground, usually a water pipe that makes considerable contact with the earth, or a special metal rod driven deep onto the ground. Without grounding, or with improper grounding, there is a great potential for people to be shocked, for fire, or for damage to appliances and motors. Most electric codes require that all receptacles, boxes, switches, and lights be grounded. While grounding wires run together with wires that carry current, grounding wires do not carry current unless the wiring system or a plugged-in appliance is damaged or defective. Not all systems require grounding. Those carried by metal conduit or cable with metal armor may not require extra grounding wires, but even these systems are grounded to components."

In the two hours before lunch, the apprentices

form teams and work on a residential wiring plan. Each project calls for a 20-amp general-purpose circuit for every 500 square feet, or a 15-amp circuit for every 375 square feet of living space. The kitchen and utility room are each to have a pair of 20-amp circuits. The kitchen is to have an outlet every four feet along the counters. Within six feet of the sink in all directions, the outlets are to be GFCI types.

In addition, the dining room, living room, and each of three bedrooms are to have one 20-amp circuit per room. Bathrooms are to be wired with GFCI outlets. None of the room lights has been included in these circuits, and several of the teams have been instructed to plan for an "all-electric" home.

As the teams divide up the work, students begin labeling sheets of paper with the names of the interior rooms they are working on. They discuss a concept known as the "demand factor" in relation to the service entry and begin to match the electrical demands for each room with wattage and materials.

As the students progress, Gene walks up and down the aisles, looking over a shoulder here, pointing out some detail there. He is distributing acetate sheets imprinted with a list of symbols agreed upon by professional architects and others in electrical contracting to denote electrical requirements. There are symbols for ceiling and wall outlets, junction boxes in various locations, single- and double-pole switches, three- and four-way switches, dimmer switches and circuit breakers. Students can see that a solid line indicates wires in a ceiling or wall and that a dotted line indicates wires in or under a floor.

When each team has finished a room-by-room list

of wiring needs, there is a brief group discussion of the plans. Next, blueprints are distributed, and each team sets about transferring their plan to the room layouts. Their finished products are not sufficiently detailed to serve as working plans from which to wire the house, but they are suitable to submit to an electrical inspector for review or approval. The final step in this project is to generate a materials list and a cost estimate for performing the work. The teams use catalogs from local electrical supply firms to compare costs and quality of needed materials. As each team finishes their project, they submit the documents they have generated and leave for a lunch break.

"Although these apprentices work under close supervision of a journeyman electrician four days a week, this classroom work is designed to make them careful, qualified problem-solvers. That's what electrical contractors tell us they need—people who can analyze a problem, see what needs to be done, and follow through on their own initiative," Gene explains.

After lunch, the apprentices return to their lab for an exercise in wiring three-way switches. This will enable a light to be controlled from more than one location. The practicality of this concept comes immediately to mind when you think about controlling a garage light from both the house and the garage, or a stairway light from both the top and bottom stair levels.

After twenty-five years as a journeyworker electrician, Gene is thinking about retiring. The union hall schedules him for four days of regular wiring and the fifth teaching this class of apprentices. He is preparing to become an electrical inspector, a solution to the problem of bending and climbing ladders all day. But he admits his reluctance to give up the teaching. "I enjoy being around people of this age, especially those who appreciate the free education they're getting and the fact that it will lead to a lifelong career with respect and good pay. There's a lot to be said for the variety involved in being an electrician, and I can honestly say I look forward to the chance to retire in a couple of years. I want to build a vacation home at a lake about an hour's drive from here, and I have three kids who will need homes of their own before long. I hope to help put up those houses and see that they are wired safely right from the start.

"Although none of my children seems to be following in my footsteps, I do have one son who's training in electronics. He's already been exposed to the 'Smart House' concept, which combines the latest in electronic technology with original wiring of a house. Maybe that will be close enough to having a second generation in the wiring business."

Not every apprenticeship program operates on the sort of schedule Gene's students follow. Official training programs are coordinated ventures between the U.S. Department of Labor, the International Brotherhood of Electrical Workers, the National Association of Home Builders, and members of the National Electrical Contractors Association, among other sponsors. All must meet the accepted number of training hours, but some programs require apprentices to work a forty-hour week and attend classes at night or on weekends.

By the end of 1999, the construction industry

alone is expected to need more than 6,000 journeyworker electricians per year. The number of apprenticeship slots is adjusted annually to ensure that sufficient newcomers enter the craft. Recruitment programs actively seek minority and women apprentices for slots across the nation. The IBEW operates a similar program in Canada.

7

Other Routes to Training

Although most of today's electricians choose formal apprenticeship as their path into the career field, a number of other programs also deserve attention, particularly if you are interested in military service or a technical school degree, or combining job training with completion of high school.

And, as we mentioned earlier, there are still jobs in the electrical field for people straight out of high school. Line installers and cable splicers, of whom more than a quarter million are at work today, represent this category.

Utility Company Jobs

Often hired as helpers or ground workers to start out, many apprentices are high school graduates with specific qualifications for these jobs. Good reading and math skills are required, since the work requires ability to understand company manuals and work orders. Many companies use aptitude tests in the hiring process, measuring an applicant's abstract reasoning ability and mechanical aptitude. A physical ability test reveals balance, coordination, and strength levels. The ability to distinguish colors is essential, since wires and cables are usually color-coded. Some knowledge of basic electrical theory may be helpful.

Other utility companies operate in-house apprenticeship programs for line installers and cable splicers. The nature of such programs varies, depending on the type of work apprentices are being prepared to do. The apprenticeship is likely to be supervised jointly by the employer and a union representing workers, either the IBEW or the Communications Workers of America. It is likely to last several years, combining on-the-job training with classroom sessions. Like the electrician's apprenticeship, the program culminates in journeyworker status.

The following is an example of the training available to a construction and maintenance lineworker for a large East Coast public utility company.

Known as "PowerLine Pro," the program is comprehensive and designed to provide line construction and maintenance apprentices with the skills and knowledge necessary to perform tasks required of a journeyworker. The program is built upon some 240 skill modules, each designed with performance-based instructional principles and methods. In layman's terms, this means that each apprentice must demonstrate mastery of a skill before moving on to the next level of training.

Self-paced, the program follows the method of individual progress through performance. At the end of a four-year apprenticeship, successful trainees have mastered all the skill modules and can prove it by hands-on demonstrations.

During the four years of apprenticeship, participants spend a total of eighteen weeks in intense training sessions at the company's central training facility. Between these sessions, experienced job supervisors oversee their efforts in the actual work environment.

A fully trained and qualified line employee could expect to earn an annual salary in the $35,000 to $40,000 range.

To find out more about the hiring policies of individual power companies, contact their personnel or human resources office.

THE JOBS CORPS

The federal Job Corps program serves several purposes. It prepares people to enter apprenticeship programs to work with electrical contractors and maintenance companies in entry-level helper positions, or to work in an electrical supply house as a counter salesperson. If you left high school before earning your diploma, the Job Corps can arrange a combination of high school and technical coursework to meet your needs.

The Job Corps training program introduces electrical theory in combination with hands-on training. Participants learn to cut, bend, and install metal conduits and wiring; hook up electrical appliances, including dishwashers, water heaters, and ovens; install electric switches and light fixtures; wire doorbells and fire alarms; repair power cords and plugs; use circuit-testing equipment; and handle power tools safely.

After completing the 800-hour training program, participants are prepared to enter an apprenticeship program or an entry-level job with an electrical contractor or a maintenance company. Program statistics show that those who complete Job Corps training earn 52 percent more than those who do not finish training.

The Job Corps program offers graduates two

distinct advantages: the skills and tools needed by a professional, and job placement assistance. To learn more about Job Corps programs, contact the state Job Service office in your community.

THE MILITARY

Enlisting in a branch of the U.S. military service is another method of training for a career as an electrician or electrical worker.

Combining job training with service to your country has proved to be a satisfying approach to career preparation for thousands of young Americans, one of whom you will read about in chapter eleven. In addition to a chance to see the world, military service gives veterans a leg up on reaching journeyworker status upon their discharge from active duty. The military pays a monthly stipend to veterans who take part in formal apprenticeship programs, and unions generally award advanced standing to veterans with electrical training in their service record.

To learn more about career opportunities in the armed forces, contact a recruiting office or consult your school guidance counselor.

The military and a number of technical schools can also train you in two of the newest frontiers for those who understand electrical theory: lasers and fiber optics.

TECHNICAL SCHOOL

A two-year technical school offers an associate degree in electrical occupations, which covers the skills and theoretical background needed for a variety of careers. Graduates can work as electrical inspectors or as electricians in electrical construction or electrical maintenance. Some graduates prefer to be self-employed, working for themselves on residential and commercial wiring.

Most community colleges and private technical schools offering this degree program have students work in a construction lab, learning about direct current and alternating current. Other courses cover such topics as basic electronics, engineering, drafting, accident prevention, electrical machinery analysis, and the use of troubleshooting and testing equipment.

To earn a degree, students must demonstrate technical skills in a variety of electrical fields, apply accepted safety standards and meet work quality standards, and demonstrate knowledge of electrical theory, math, and physics as they apply to construction.

Students also learn to use and care for electrical tools and materials. They learn to read and develop blueprints to perform installations that comply with the National Electrical Code. They generally learn to set up ladder relay logic systems and convert them to electronic programmable control systems. The same is true of basic electronic control circuitry, devices, and schematics. They also learn to interpret ideas and develop plans through communicating with others.

Recommended high school subjects include one year each of general math, basic algebra, and science; a year of advanced algebra is desirable.

Another area of keen interest in today's technical schools is the associate degree in electronics technology with its emphasis on laser electro-optics. Such training focuses on specialized courses in physical sciences and laser optic components and systems. A basic understanding of electrical theory is required, built on fundamental coursework and lab experience in DC and AC electric circuits, solid state devices, and digital devices and systems.

Interested? Perhaps you'd like to see the course objectives for one of these degree programs:

1. apply working knowledge of ac and dc circuits
2. understand the theory and operation of solid state devices, linear and digital integrated circuits, and microprocessors
3. solve mathematical problems relating to circuit analysis, digital electronics, and other systems
4. read and interpret technical literature and specifications
5. communicate verbally with others and write presentable technical reports
6. perform accurate and valid parameter measurements with laboratory test instruments while observing standard safety practices
7. program microprocessor-based systems and interface peripheral devices
8. demonstrate knowledge of the properties and propagation of light
9. apply the laws of reflection and refraction to light as it passes through an optical system
10. demonstrate knowledge of optical equipment and hardware and its applications
11. describe the generation of light in a laser
12. describe the different types of lasers and their characteristics

13. classify lasers according to their characteristics and applications
14. practice laser safety procedures and precautions
15. demonstrate knowledge of the theory and operation of laser support equipment
16. demonstrate knowledge of the theory of operation and use of laser power and energy measurement instruments
17. demonstrate knowledge of and have experience with laser system applications

A similar high-tech program prepares people for careers using fiber optic technology. After two years of study, participants earn an associate degree in electronics technology with a fiber optic/communications emphasis. Again, core courses focus on basic knowledge and laboratory experience with DC and AC electric circuits, solid state devices, and digital devices and systems.

Students specialize in laser and fiber optic devices and systems. The degree prepares them for jobs as telecommunications technician, technical sales consultant, broadcast engineer, cellular phone technician, microwave technician, and a variety of other technical positions.

Admissions representatives of technical schools visit high schools to meet with interested students. Many also produce videotape "tours" that tell the story of their programs and courses. Once your school guidance counselor knows of your interest, it should be easy to find a technical school that meets your needs.

8

Preparing for the Professional World

How will you actually know whether the electrician's trade is right for you? It has many attractions, and yet you may not like some aspects of the job. It is better to learn as much as possible about a career before committing to training. If you find that the advantages outweigh the disadvantages, you are far more likely to be satisfied with your career choice over the three or four decades you will spend working.

Points to Ponder

First, consider what attracts you to the electrical trades. Could it be the challenge of controlling something as powerful as electric current? Perhaps it is natural curiosity about electrical theory, or the scientific nature of the formulas and equipment. It could be the relative freedom with which electricians conduct their workday, or the chance to be paid while learning valuable job skills. There is much to be said, too, for the feeling of accomplishment that comes from performing important work and doing it well.

Knowing about an occupation is just half the assignment, however. Next you must gauge how well-suited you are to carrying out the assignment. To help in this process, ask yourself some realistic questions.

For example, how is your general health? As a group, blue-collar workers are healthier than the general population, and that's a good thing, considering the strenuous tasks they are often asked to perform.

Would you be willing to take a complete physical exam? Most hirings depend on the results of such an exam. If you are hesitant, it will reduce your chance of being hired.

Do you have access to reliable transportation? Construction electricians in particular may be required to travel to job sites fifty to 100 miles away.

Also of concern to the construction electrician is the ability to budget money for a rainy day. In the early years, seasonal layoffs are to be expected. In addition, construction projects are often held up by bad weather and delays in receiving materials, which can even shut down a job site. Unemployment benefits and SUB pay certainly help, but they do not totally replace wages.

Utility workers may also face weather-related slowdowns, but more likely they will be expected to work outdoors in adverse conditions. Climbing ladders, working on or near electrified lines, carrying materials or tools—those are all part of the game.

Next, consider how you relate to others. Are you a team player? Remember the safety hazards that accompany this work. At times your actions will determine your own safety and that of your coworkers. During your personal interview, on-the-job training, and probationary periods, you will be constantly assessed in this vital area.

You are also likely to face a few informal "tests" devised by your coworkers. As is the case in many fields of work, newcomers are often subjected to pranks. The best response is a sense of humor. While you are low in seniority, you probably will be given the worst jobs; with seniority, you will be able to bid for jobs that pay better and are usually easier to do.

Two personal qualities that will mark you for success early in your career are a high level of self-confidence and the ability to stick with a project.

APPLYING FOR A JOB

A union journeyworker's card is the surest guarantee of work in the electrical trades. No one receives such a card unless he or she has demonstrated skill and gained considerable on-the-job experience. Aside from union hirings, the electrical industry relies on two standard employment tools: the job application and the personal interview.

The Résumé

The résumé, a formal history of your educational and employment background, is not a requirement for job-hunting in this field. Still, much of this information will be requested on the application, and you may find it easier to complete the application by referring to such a chronological list. Remember to keep a record of the time during which you attended each school or held each job. The complete address and phone number of all schools and employers may also be needed.

Sample Résumés

While the names have been removed to protect the privacy of individuals, the following résumés are good examples of the sort of training and work experience electricians bring to the job market today. It is important to note that job turnover is quite low among professional electricians. Generally, a desire to live in a different region or to branch out into a different area of the career field is the motivating factor when an electrician decides to seek other employment.

Résumé Number One

Position sought: Supervisor of electricians

Relevant experience:
August 1995 to present, electrical subcontractor to Mountain Vista Homes subdivision. Performed rough and finish wiring to meet NECA Code standards. Ordered materials, scheduled work for four-person crew, maintained records, and coordinated electrical inspections.

April 1990 to July 1995, inside electrician, Middlebury Mobile Homes, Inc. Performed residential wiring at mobile home factory, including wiring electric furnaces and laundry and kitchen appliances.

March 1987 to February 1990, member of U.S. Navy; electrician's mate aboard aircraft carrier USS *Enterprise*. Graduate of Navy technical training center. Responsible for repair and maintenance of electrical systems, including electronic communications equipment. Honorably discharged following four-year enlistment.

Education and training:
1986 graduate, Mountain View Vocational-Technical High School, with a certificate of completion in

construction trades/electrical. Successfully completed intensive twelve-week course, U.S. Naval Training Center, Great Lakes, Michigan.

Completed series of on-the-job training modules during carrier assignment equivalent to electrician's apprenticeship.

Passed state electrician's licensing exam, February 1990. Scored in top 20 percent of examination group.

Completed all necessary continuing education requirements while maintaining state license.

Certifications:
Volunteered for instruction in nuclear operations and safety/rescue topics while aboard the *Enterprise.*

Hold valid journeyworker electrician's card; also have U.S. government security clearance.

Special interests, talents:
Building and flying remote-controlled aircraft. Have computer skills including word processing and database management. Enjoy rock climbing. No fear of heights or confined spaces.

Professional references available upon request.

Résumé Number Two

Position sought: Journeyworker Electrician

Purpose in relocating:
As the parent of a four-year-old with asthma, I am seeking employment in the Southwest for my child's health and physical welfare.

Education, training, and experience:
In 1996, successfully completed an IBEW-sponsored apprenticeship consisting of classroom instruction and supervised on-the-job training. (Received advanced placement due to technical school degree.) Assigned to industrial and commercial wiring duties while apprenticing with the Allen Company Electrical Contractors, Inc., for nearly three years.

Hold an associate degree in electrical occupations, Pennsylvania Technical College, Williamsport, Pennsylvania. Maintained an overall grade point average of 3.0, with a higher average in major course work. Also employed part-time during college at a building supply center.

Trained to meet the standards of the National Electric Code. Skilled in the use of testing equipment and programmable control systems.

1991 graduate, Central High School, with courses in shop math, algebra I and II, plane geometry, wood and metal shop, basic computer literacy, and cooperative learning.

From age sixteen until I started trade school, I worked as a part-time counter person at the Home Center building supply store in Summerville, Pennsylvania. Duties included waiting on customers, cutting glass, repairing windows and screens, loading and delivering materials, mixing paint and stain colors, generating computerized receipts for charge purchases, taking inventory, and related activities.

Personal and professional references available upon request.

The Job Application

It is a good idea to read the entire application before making any marks on it. Crossed-out lines, mistakes, and erasures reflect negatively on you. Complete each item fully and truthfully, assuming that the information will be verified. If you have any questions, ask for clarification before you attempt to write an answer. Sign the form if requested to do so. Also, check the back of the form for additional questions.

Following is a sample application for a position with a large public utility company that includes a nuclear power generating station. The first page, headed "Representations, Authorizations, Conditions, and Release," asks that the applicant carefully read a series of seven statements before completing the application:

The Company is an Affirmative Action and Equal Opportunity Employer, complying with federal, state, and local regulations. The Company does not discriminate in hiring or personnel practices.

Employees are hired on an at-will basis, and employment can be terminated without cause. No contract of employment is in effect covering a definite term or conditions of employment.

Candidates for employment authorize the Company and/or its agents to conduct an investigation to verify the truthfulness of statements made on the application and during interviews. A report may be prepared, including personal interviews with the applicant's neighbors, friends, or other acquaintances, regarding the applicant's character, general reputation, personal characteristics, and mode of living.

Any employment offer is contingent upon successfully

completing a medical exam and the background investigation described in part C above, along with satisfactory compliance with the requirements of The Immigration Reform and Control Act of 1986. There will be testing for the presence of drugs and/or alcohol. Certain positions may also require a psychological evaluation and fingerprinting for purposes of obtaining an FBI Criminal History Report (the presence of a criminal record is not necessarily a bar to employment).

Compensation and terms of employment for jobs covered under a collective bargaining agreement will be subject to its terms. Any deliberate misrepresentation or omission of a fact in my application or in any interviews in connections therewith may be justification for refusal of employment or termination of employment.

By my signature below, I affirm that I have read and understand the representations, authorizations, conditions, and releases made or given above.

The second page, entitled "General Information," asks for the applicant's full name, address, telephone number, and Social Security number. Next, the applicant is asked to choose from nineteen job categories, listing three choices in order of preference. In addition to office, security, and maintenance/mechanical jobs, the application includes positions as an electrician and in electrical construction and maintenance as well as power plant generation.

Applicants are asked whether they would accept an assignment that required shift work, including a rotating shift. They are also asked whether they have ever worked for one of the parent company's six affiliate operations.

The next page asks applicants whether they have relatives presently working for the utility. The next question is, "Have you ever been convicted of a crime which has not been expunged or sealed by the court?" If the answer is yes, the applicant is asked to explain. The next line states, "Please note that the presence of a criminal record is not necessarily a bar to employment."

The following section provides a chart in which the applicant is asked to recount his or her education and training, including high school or vocational schooling, college, and other training experience. It calls for details including the city and state where the institution is located, the highest level completed, major course of study, and degree or diploma received. In addition, space is included to list any professional certificates or licenses earned.

Page five is marked "confidential" and includes a small boxed section, "Internal use only." In addition to blanks for name, Social Security number, and position being applied for, a section headed "Voluntary Self-Identification" notes that certain information is sought for record-keeping in compliance with federal and state laws. Responses are voluntary and will not affect the employment decision. In addition, the information will be kept confidential in a file separate from the job application.

What follows is a checklist of sources with the question, "Through what source were you referred to our Company?" The choices are advertisement, walk-in, recruitment, one of the system companies, mail submittal, minority agency, state employment agency, employment agency, college placement office, other (specify), employee referral (name).

Next are check-off boxes to indicate gender (male or female); race/ethnic data (white, black, Hispanic, Asian or Pacific Islander, American Indian, Alaskan native, etc.); veteran status (disabled veteran or Vietnam-era veteran), and disability status. If the applicant answers yes to the disabled individual or disabled veteran question, the form asks whether the applicant possesses any special methods, skills, or procedures that might qualify him or her for positions otherwise impossible because of the disability. Another question asks what accommodations could be made to enable the applicant to perform the job properly and safely.

THE JOB INTERVIEW

Most hiring decisions rely heavily on the results of a job interview. This involves one or more face-to-face meetings between the candidate (you) and someone who represents the company doing the hiring. Although it is natural to feel a little nervous before a job interview, there are ways to prepare yourself for the experience, thereby boosting your self-confidence and helping you make the best impression on the interviewer.

Start by knowing about the job you want and the company in general. If you know someone who already works for the company, ask questions. Before your interview date, stop by the office reception area and ask for copies of company brochures or annual reports. Why do advance research? Because during the interview you should ask some questions, too. They should indicate that you want the job enough to take an interest in the company before you are hired. Your questions should also reflect initiative, another positive trait. If you sit in silence or fumble for an answer when the interviewer asks a question, you will probably lose points.

Asking those questions is somewhat difficult. For that reason, you should practice them in advance so that you sound relaxed and in control of your thoughts during the interview. What topics are appropriate? You can't go wrong if you ask about:

- The job duties, working conditions, and the opportunities for advancement.
- Qualifications wanted: physical, emotional, and educational.
- Does the job include fringe benefits?
- What are the working hours?
- Will you work alone or in a group? How is the group organized? To whom would you report?
- When will you know if you've been hired? What are your next steps?

Another way to feel more comfortable during the session is to prepare for the interviewer's questions. It is possible to anticipate some of them. For instance, many interviewers ask:

- Why do you want to work for our company?
- Why do you think you are qualified for this job?
- Tell me about yourself and the work you like to do.

Later in the interview, you might be asked about your family, how you spend your spare time, what you think you might be doing in five years, what you consider to be your major strengths and weaknesses, the subjects you liked best in school.

Put your thoughts on paper, and while you're at it, consider a few more commonly asked questions: What did you do on your last job? What makes you think you are ready for this job? What are your best successes and accomplishments? What made you leave your last job? How much money do you feel you should be earning? Why should we hire you instead of someone else? How long do you intend to stay with us? What would you like to tell me that is not on your job application?

Practice a response for each question until you feel comfortable with it. Be sure your response is genuine and shows something of who you really are. All the while, remember to emphasize your strengths; don't forget to admit, yet downplay, your weaknesses.

The Big Day

To give yourself an edge, be neat, clean, and alert. Plan to get a good night's sleep, and set the alarm a little early. Wear appropriate clothes. For a blue-collar job, choose clean, presentable jeans and a dress shirt. A recent haircut is advised. Don't smoke in the reception room or during the interview, even if invited to do so. If you have a completed application, make sure your name, address, and phone number appear on every page. Place the unfolded pages in a large envelope, or clip them securely inside a folder.

Don't be late; in fact, try to arrive five to ten minutes early. Tell the receptionist your name and whom your appointment is with; then take a seat and try to appear calm. Don't shuffle through your paperwork or do anything unusual; the receptionist may have been asked to report on the behavior of job candidates.

You may be asked to take one or more tests before the interview begins. Electrical workers are frequently given tests that measure mechanical aptitudes or math skills. Remember that people who are tired or don't feel well generally do poorly on tests. Sometimes being a little nervous actually helps focus your attention. Just read or listen carefully and follow the directions.

In today's competitive market, many job-seekers face a few turndowns before scoring a success. By taking time to prepare for your interview, you will certainly improve your chances of being hired. Remember that today's jobs demand workers with initiative, people who can solve problems and follow directions. Your preinterview work will sharpen skills in each of these vital areas.

Remember to look the interviewer in the eye, which is not to say that you should stare. Lean forward a little bit and demonstrate that you are a careful listener. Don't fidget. An occasional gesture is fine; otherwise keep your hands folded in your lap or resting quietly on the arm of your chair. Your voice should be clear and assertive in tone. As with gestures, be conservative with your smile, yet don't go through the whole interview with a poker face. Above all, don't fall into the trap (sometimes deliberately set) of complaining about past employers.

Only the biggest corporations can afford full-time professional interviewers. Remember that the person you talk with may be as uncomfortable in the interview situation as you are.

Most interviews end with a noncommittal comment such as, "Thanks for coming in, we'll get back to you."

You may leave with no idea how you did or whether you have a shot at the job.

Should you simply wait for a phone call that may never come? Recruitment counselors recommend that you send the interviewer a short follow-up letter within a day or two, thanking the person for seeing you and mentioning again your interest in the job.

A week or two later, it is permissible to make a follow-up call, asking the interviewer about the status of the job. Has it been filled? Are you still being considered? If not, it's time to apply elsewhere. Even if you don't get the first job you apply for, taking part in the interview is a valuable experience that will help you relax and feel more confident the next time.

9

Opportunities to Specialize

For many electricians, the wage increases that seniority and new contracts bring are sufficient reward for their time and effort. Still, being assigned supervisory or management duties can lead to greater income and job satisfaction. The journeyworker electrician has many opportunities to develop special talents within the field. A lineworker may switch to electronics or supervision; a maintenance electrician may specialize in instrumentation. Because electricians are often naturally curious and highly motivated, they are generally eager to take part in advanced training. The pay scale is generally $0.50 to $2 per hour more than an experienced journeyworker's wages.

Here, in some detail, are descriptions of several career specialties that an electrician may seek:

Control Specialist

Person who services and repairs control equipment and devices used to start, accelerate, regulate the speed of, brake, and stop electrical machinery. This specialist needs to be familiar with electrical theory and application, but must also have experience with hydraulics, mechanics, electronics, and pneumatics. Problem-solving skills are required to work with complex industrial production equipment. Test equipment used includes voltmeters, ammeters, wattmeters, oscilloscopes, signal generators, capacitor checkers, resistance bridges, tachometers, frequency meters, and logic probes. These expert troubleshooters know the circuitry of many devices.

ELECTRICAL INSPECTOR

This specialist provides a service to municipalities and the electrical construction industry as well as to consumers. Because of the hazardous nature of electricity, many precautions must be take when installing electrical equipment and systems in homes and other structures. Safety is the cornerstone of various building and electrical codes, and the electrical inspector has the job of ensuring that such safe procedures were followed. The work is usually guided by written codes, particularly the National Electrical Code. An electrical inspector must have both practical and technical knowledge of the field, plus the ability to communicate effectively with electric power company personnel, electrical contractors, and their supervisory employees.

These careerists conduct field inspections, work from checklists, file reports, and issue permits or approval orders. They must be conscientious workers, since inattention to duty may result in fire or electrical shock.

Most electrical inspectors are employed by municipalities, but others work on a contract basis. They may earn from $1,800 to $3,000 per month. The job usually requires the use of a car or truck.

ELECTRICAL CONTRACTOR

This is a rewarding and promising field for those with entrepreneurial skills and a knowledge of the trade.

Contractors and their key employees draw on their experience to build a business for the future. Many started out as journeyworkers and may have worked their way up, spending time as supervisor, job estimator, or operations manager before going out on their own. They gained experience in making decisions about planning, scheduling, and controlling a business operation. They can rely on support by the National Electrical Contractors Association as they launch their business venture.

Today, half of the major electrical contractors in the United States are college graduates. Since the work is becoming increasingly complex, this trend is likely to continue. Typically, the degree is in electrical engineering, although business degrees are also common. To succeed, an electrical contractor needs a combination of education and experience, initiative, foresight, good judgment, personality, and communications skills.

Electronics Journeyworker

This specialty has become closely allied with the field of electricity in the past several decades. Most electricians now need to know basic electronics, and many who seek a challenge decide to specialize in this field. Usually workers who install, service, and repair electrical/electronic equipment enjoy favorable working conditions and diversified duties. Whether the task is to solder a wire to a terminal or build and maintain complex switching devices, the electronics journeyworker uses a combination of electronics and electrical know-how. Particular skills as a troubleshooter are required, since much of the work involves locating and correcting circuit faults. This worker must be familiar with sophisticated test equipment, signal generators, power supplies, electronic switching devices, and similar equipment.

ENGINEER

Another way to specialize as an electrical worker is by becoming an engineer. The high standards set by electrical apprenticeship programs, along with the rise in the general level of education, provide a good base for pursuing this option. Engineering technicians study for two years; an engineering degree requires four years. Both degrees generally concentrate on a specific area, for example, electric power, electronics, electrical equipment manufacturing, communications, electrical maintenance, or engineering consulting.

INSTRUMENTATION TECHNICIAN

This specialty deals with intricate instrumentation and control systems that require installation, service, and repair. A technical background is needed, and the work usually is done in an industrial setting where the equipment worked on is designed to process chemicals, control the movement of ships and aircraft, produce electricity or petroleum products, manufacture metals, or guide other industrial processes. Electricity, electronics, pneumatics, and hydraulics are commonly involved in this challenging field. Technicians are frequently required to work on small components in cramped spaces, making minute adjustments. The ability to interpret and work from complex circuit drawings is needed, along with experience in the use of sophisticated test instruments. Problem-solving skills are also required. The military service has trained many of today's experts in this field, and veterans have little trouble finding good jobs when they return to civilian status.

This specialty is interwoven into the country's technological future, making opportunities virtually unlimited. Most industrial plants are moving to process control systems in the never-ending search for greater productivity.

JOB ESTIMATOR

This specialty involves duties that are both interesting and challenging. Every contracting business has at least one person who performs job estimates. Sometimes it is the business owner, but a supervisor or skilled specialist is likely to be given the responsibility. Good estimating skills require a combination of engineering knowledge, skill, and practical experience, plus the ability to use the tools and techniques of the estimating trade. These include tables, charts, and graphs related to time and materials.

To carry out these duties, one must be able to understand blueprints and job specifications. An estimator should also have a good eye, prudent judgment, and ability to anticipate time requirements and spot potential delays or on-site construction changes that could significantly affect the job to be done. The best estimators keep up-to-date on the latest tools and materials and their cost and availability by reading trade publications. An estimator's day is a combination of office work and site visits or meetings with architects and engineers. When business is brisk, an estimator may put in long hours and face job stress. Many estimators participate in profit-sharing plans as part of their compensation package.

SUPERVISOR

Supervisors are needed in all areas of the electrical industry: construction, maintenance, and power utility. They are leaders in a particular job area, chosen for their mechanical and problem-solving abilities, initiative, and communication skills.

Duties include making decisions, giving instructions, evaluating the performance of workers, coordinating job details, and keeping records. The supervisor is also responsible for the safety and well-being of workers. When job training is required, the supervisor often provides it, or may supervise apprentices. Overall, it is the supervisor's job to see that work is performed within budget limits and with a degree of quality.

Job titles include electrical construction supervisor, line crew chief, and shop supervisor. This specialist usually serves as a mid-level communicator between workers and upper management. Communication skills are also required to work with engineers, architects, electrical inspectors, safety personnel, and the public.

The supervisor must be familiar with local and national electrical codes and safety practices. Pay scales are from $0.50 to over $3 per hour more than the workers being supervised.

TEACHER

Instructors in vocational-technical schools and community colleges are often recruited from commercial or industrial backgrounds. The selection process takes into consideration professional training, experience, educational background, technical knowledge, professional attitude, and personality.

In selecting personnel for technical assignments,

greater emphasis is placed on professional background than on academic accomplishments. For this reason, craftspeople and skilled technicians capable of imparting knowledge and nurturing the skills of others have a good chance to move into instructional positions. Instructors must know their field thoroughly and be able to gear their presentation to students of varied learning abilities. Careers in teaching are especially rewarding for those who enjoy working with other people. The school calendar also means that these jobs include extended time off with pay.

10

Becoming Your Own Boss

Becoming "the boss" is the dream of many American workers and one that could certainly become a reality for a qualified electrician. An electrician with an interest in contracting or service work could build a successful business venture by applying management skill, a little investment capital, and a lot of energy.

A word we hear often is "entrepreneur," borrowed from the French and meaning someone who takes the risk of organizing and running a business in the hope of making a profit. Because they are careful problem-solvers, electricians often make successful entrepreneurs.

Hard-working people make the transition from employee to employer for many reasons. The freedom to make decisions and explore opportunities is often the motivating factor. Some people go on their own because they believe it is the only way to test their talents and potential. Others frankly admit that they enjoy the sense of power and control that ownership brings. Some entrepreneurs liken the experience to "growing" a business, much as a gardener raises plants from seed to harvest.

Although fewer than 10 percent of today's electricians are self-employed, that figure is likely to change as more homebuilders, general contractors, municipal governments, and other groups call upon trade specialists to perform needed work on a contract basis.

The state in which you launch your business may set some requirements that you must meet, including providing worker's compensation insurance or other coverage in case of accidental injury on the job. These requirements should be no obstacle to a fully qualified journeyworker. The key to a successful start in business is careful planning, something an electrician does as a matter of habit. The best time to start your own business is after gaining some experience in supervising workers and managing time and materials for someone else.

Because cash flow problems can strangle a start-up venture, you must be prepared to live on your savings until the volume of business expands. Even if your efforts make money from the start, you may need to pay for things like materials and equipment, legal advice, or advertising. Good financial planning will protect both your business and personal credit rating, another essential of a successful business.

Electrical contractors bid competitively with other businesses for available work. In large building projects, such as hospital additions, new school buildings, shopping centers, and office buildings, a general contractor is selected to perform specific parts of the work, such as erecting the building's envelope and interior spaces. Subcontractors bid to install the "systems," including plumbing, heating, and electrical wiring.

These projects are on a large scale, requiring detailed blueprints prepared by an architect. Complex bid documents and contracts govern the work, which may take many months to accomplish.

Electrical service companies are generally organized on a smaller scale and compete for clients who need wiring installation or repairs to their property. The work is smaller in scale and usually does not require architectural blueprints. Instead of a contract, the project is described in a simpler document known as a work order. Some service projects are bid; the customer pays an hourly charge for other work.

In both types of business, experience in estimating time, materials, and labor helps the owner balance price and profit factors.

If you start an electrical contracting firm, you may be able to bid work for your former employer, especially if you worked for a general contractor. Many construction electricians who wired houses for a living take this route as entrepreneurs. Advertising, membership in the National Association of Home Builders, and word of mouth will help fledgling service companies gain business.

In the beginning, the owner of a new business works long hours, at the job site by day and on bids, bookkeeping, or material orders in the evening. Small contractors may trace their start to a single successful bid. It is important to keep up a steady pace, adjusting to growth when it happens and downsizing if the economy slows. A successful contractor continues to bid jobs in a size category he or she can handle, moving up with experience as a guide.

Marketing

Most contracting firms and specialty services rise or fall on the strength of their reputation. At the job site, your employees represent you and your firm's reputation.

They must understand that the future of your business depends on the attitude they convey and the quality of their work.

To promote the business, some electrical contractors rent exhibit space at home shows or set up a sign with the company logo at the work site, particularly if there is a grand opening or open-house event. Display ads in telephone directories, listings in Chamber of Commerce promotional materials, or ads in the local media are other options. Still, word-of-mouth reports of the excellent work you do is the best and most cost-effective form of advertising. You will build a client base for repeat busines and also gain valuable referrals.

As you take the first steps in your electrical career, study the success of others. You will inevitably learn that "luck" has little to do with it. Experience, technical knowledge, drive, ambition, patience, and the ability to solve problems—these are traits successful business owners share.

PERSONNEL

You will find yourself wearing many hats, at least in the early years of operating your business. Electricity can be dangerous, and you will need to be safety-oriented, as will your employees. It will be your responsibility to enforce employment laws and communicate regulations clearly to workers. You will need the ability to judge people's character in order to make hiring and firing decisions, assign jobs, and protect the good name of your business. Your ability to encourage and motivate workers will also be tested.

There is also the business office to be run. Someone

must handle the phone, written communications, and bookkeeping. Tax reports and compliance documents will also be required should you bid government contract work.

It is for you? Entrepreneurs generally end up working longer hours than their employees. Are you willing to invest the time and take the risks? Luckily, in the contracting business you don't have to make a lifelong commitment.

SUPPORT NETWORK

One way to increase your chances for success is to benefit from the experience of someone who has been there. A mentor is someone who has experience in your area of concern, someone who can advise you and teach the specialized skills of your business. SCORE, the Service Corps of Retired Executives, operates throughout the United States as a sort of "big brother" organization for fledgling business owners.

If you are a woman launching an electrical contracting business, you can find support through the National Association of Women in Construction. With more than 226 chapters in the United States and Canada, this organization represents nearly 9,000 women employed in all phases of the industry. NAWIC offers networking resources on the latest in such areas as construction techniques, women's business ownership, and office management. The group also sponsors educational programs and regional conferences and helps to establish construction contacts. Financial support for education is offered through a scholarship program. NAWIC is based in Fort Worth, Texas, and has a toll-free number: (800) 552-3506.

Most state governments now offer business service centers with programs and information of value to entrepreneurs. Local Chambers of Commerce and Industry also organize workshops to assist small-business owners.

11

A Day in the Life of an Electrical Worker

There is no way better to learn what an electrical worker does on the job than to "shadow" that person—that is, to go to work with him or her—to discover what a typical day brings. This section is a series of vignettes or word pictures depicting a typical day in the work life of several electricians, from apprentices to experienced journeyworkers.

Some are union members; others work independently. The older, more skilled electricians are called journeyworkers, and they often supervise an apprentice or two while they go about their own day's work. The number of electricians at a construction site depends on the size of the job. For example, wiring a new home rarely requires more than two workers, whereas the average commercial job requires five to eight electricians, and a large industrial job may provide employment for more than fifty electricians.

Just over half of the 500,000 electricians employed in the United States work in the construction industry. Contractors and construction firms are the leading employers. Maintenance electricians are likewise employed by a manufacturer or business. In fact, fewer than 10 percent of all electricians are self-employed.

Being an electrician brings the opportunity to work

on varied projects, to travel, and to train others in a craft that has a long and proud tradition in the American workforce. These ideas are held by Dave, an IBEW electrician from suburban Pennsylvania, who has worked in Las Vegas as well as his home area over the past several years.

"I've had my union card for over twenty years, and now that my children are grown, I've started taking advantage of an urge to see the country, signing on for construction work in states that interest me. Through the week, I'm bending and pulling wire just like always, but on weekends and holidays, I get to see places I've never been to before.

"Most of last year, my wife and I stayed in Las Vegas, Nevada, where I worked on a resort hotel complex. It's a high-rise hotel, a gambling casino, and a tourist attraction all rolled into one, with the erupting volcano out front as its major feature. As a of matter fact, I helped to wire the controls that operate the volcano.

"We rented a condo in Las Vegas and considered that entire time a sort of working vacation. My schedule was basically 7:30 AM until 4 PM, so after a shower and a short rest, the evenings were free. Often, we'd go to dinner and take in one of the early nightclub shows. There's so much to see and do in Vegas, and it's become a family-oriented town. On weekends, we'd drive over to California or down to Arizona. You could be in either state in a few hours. We saw the Grand Canyon several times, plus Indian cliff dwellings.

"During the week, my workplace was familiar and reliable, the same tools and blueprints, safety codes and equipment as always, but after work, I had new worlds to explore. And I knew if for any reason I had to head back home to Pennsylvania, I could do that, but I didn't mind missing one winter.

"The lights on the strip in Las Vegas are fantastic. It's the perfect place for an electrician specializing in neon signs. Along with all the commercial work, Chester County, Nevada, has a residential building boom going on. If you'd rather wire houses, there's plenty of that work."

While Dave and his wife always intended to go back home to Pennsylvania, it was the lure of a fascinating job that sent him East again.

"My IBEW Local got the call to take on a big job at a military base near my home. The base is building several new flight simulators for military pilots to train in. I've always had an interest in aviation, and the chance to work in that environment was too interesting to pass up. So that's what I'm doing now, Monday through Thursday. . . . I work four ten-hour shifts. After a couple of weeks, the long weekends got to me, so I've been helping a friend with a remodeling business. He has a lot of senior citizen clients, and I get a good feeling from helping those folks. You know, making sure the wiring in their homes is up to code and safe from fire hazards. I usually work three or four hours on Friday afternoon or Saturday morning. Not that I need the money, but the residential work is a refreshing change from the heavy wire we pull for the government project. I enjoy meeting people as you do in residential work."

Like many electricians, Dave confides he was fascinated by electricity from an early age. "I had all those kits you used to build batteries, doorbells, incandescent lamps and radios. I was fascinated by my mom's vacuum cleaner from about the age of two on. Then it was radios and record players. I enjoyed shop and mechanical drawing in high school. Since you need to read blueprints, that was a good start."

Asked if he'd recommend the career of electrician to others, Dave was quick to assent. "I can't think of many other jobs where you have the flexibility and good income an electrician can command."

Chad

Chad is a registered electrician who works for a plumbing, heating, and cooling contracting firm that also installs custom kitchens and baths. He enjoys the variety of the work he does, and he has also developed extensive skills as a carpenter. Regardless of the nature of the assignment, however, safety comes first with him.

"Some strange and dangerous things can happen when electricity is involved," he pointed out, easing into story-telling mode. "You've got to remember that electricity flows, and you can get hurt making contact with household current.

"We had a call at the shop from some long-time customers who had just bought an old plank house that had been updated with aluminum siding. The odd thing was the reaction their dog kept having to this house. If the dog was outside, he was afraid to come indoors, no matter how the owners coaxed him. At their former home, the dog had developed the habit of bumping against the storm door to signal that he wanted to come in. But at this country house, he shied away from the door.

"The people suspected a problem with this aluminum storm door, because they were also getting a little tingle from touching the doorknob, especially if it had rained.

"Well, all this sounded a bit suspicious to me, so I took my meter out there. Sure enough, the doorknob was hot, and there were places where the aluminum siding was registering 120 volts. The people touching that doorknob had shoes on, while their poor little dog had four damp feet on the ground and a wet nose that would touch the door or siding. As you know, aluminum is a great conductor of electricity. Apparently when the siding was nailed to the plank walls, someone pierced the electrical wiring with a nail.

"It took four or five hours to locate the area where the wire had been pierced, but the alternative would have been simply to shut down electric power to a portion of the building. It's just lucky no one was seriously hurt before the dog brought it to the owners' attention. That was one of the strangest service calls I ever went out on."

CATHY

As a single mother of two, Cathy spent seven years creating a path to success. "If you have a trade, no one can take that away from you. Never settle for minimum wage, and never feel stuck. Success is not as difficult as it seems," she advises.

In search of financial security, opportunity, benefits, medical insurance, and a retirement plan, Cathy used curiosity about the electrical industry to launch a new career path that led first to a union apprenticeship program. She was the only woman in her classes for most of those years and admits there were times she questioned whether she could really see it through. Not only did she earn a journeyworker's card, with two years' experience under her belt, she launched her own contracting business.

"I'm always learning something new, and the fact that I deal with technology and understand it, that's rewarding, too." While much of her company's work involves updating electrical systems in office environments, she wants young women to know there is a broad range of jobs in electrical work. "Lighting consultant, fiber optics, automation and control work, even engineering...." Her advice? "Sign up for science and math, and see your guidance counselor early. Two years or technical school, or an apprenticeship that combines paid work with training may be just what you need."

RON

Ron begins the day at 8 AM in the dispatcher's office of an electric service company. He joins six fellow electricians, coffee mugs in hand, to receive the day's assignments. Each worker has a service vehicle, either a light pickup truck or a utility van, stocked with tools and materials of the electrician's trade. Some of the jobs being assigned will take only an hour or so to perform, so the worker receives a list of five, six, or more projects; others are continuations of more involved projects begun the day before.

"There's a lot of variety in this work, and that's why I like it. It's one of the first things I picked up on as an apprentice, and something I recommend to people who think they'd like to become an electrician. The basic theories apply to many kinds of wiring jobs; only the circumstances change. You can rely on your skills, yet know the job won't get boring—or if it does, at least it will be finished soon."

As a union electrician, Ron is a member of the

IBEW. "I did my apprenticeship in Pennsylvania, but after working for ten years I took advantage of union job placement and moved my family to South Carolina. The winters here are more to my liking. The company I used to work for had a maintenance contract on electronically controlled traffic signals in two neighboring towns. When those systems malfunctioned in the winter time, it was no fun working from a cherrypicker bucket in the wind and cold, not to mention the traffic buzzing by."

In the past six months, Ron has directed a four-person crew wiring an addition to a nursing home; installed dozens of dishwashers, hot water heaters, and similar electric home appliances; and installed underwater lighting for several swimming pools.

Whereas some union members work strictly in construction and change assignments or even employers every few months, service electricians work for a single employer, usually a small-business operator who bids on contract work or charges an hourly rate. Customers include residential, business, commercial, and sometimes industrial property owners.

"If you enjoy solving math and science problems, then understanding basic electrical theory should be a challenge you can succeed at," Ron points out. "Another basic requirement is that you can identify colors. The color-blind electrician has a real problem, since wiring manufacturers use a system of color-coded wires that must be matched up and wired together.

"In service work, it's important to have a good sense of direction. Handling a series of assignments in one day in different parts of the city requires that you get from one address to the next without wasting time in traffic. There's not a whole lot of weight-lifting involved, though you will be moving ladders in and out of the vehicle and carrying a toolbox and materials. None of these things is really heavy, though. Sometimes the work requires standing or kneeling in a small space.

"Because of the nature of residential wiring, you often work on the job twice, with some time between sessions. The rough wiring comes first, before the walls are built, and then you go back toward the end of the job to do the finish work—wire the outlets and install light fixtures or appliances, that sort of thing."

Jeff

Jeff learned his trade in the U.S. Navy. Assignment to an aircraft for more than three years taught him to work in confined spaces and to do things right.

Within weeks of his discharge from the service, Jeff had taken the entrance exam given by the IBEW and received a journeyworker's card. The transition was relatively easy, since all electricians are trained to follow National Electrical Code specifications, whether they learn the trade in the military or through apprenticeship. With help from the union hall, Jeff has located a job with a home builder.

"After being at sea so long, I'm ready to settle down and start a life for myself on dry land. Of course, in construction you have to take into account the seasonal nature of the work. I learned how to save money in the Navy, so I'm not worried about the chance of a winter layoff. I can just work some overtime and save that money for later, or tighten the purse strings for a while—no problem.

"Two of my uncles are electricians, and my family has been in commercial contracting for years, so building sites are nothing new to me. The real surprise was when I met a Navy recruiter at our high school career day and discovered I could learn the trade while seeing the world. After basics, I drew San Diego as a home port, but aircraft carriers are deployed much of the time. We had a cruise to the Mediterranean and spent several months in training exercises in the Pacific. I got to see the world all right—Rome and Greece, the Philippines, and Hawaii.

"Some of my buddies got jobs at the shipyards, but I had had enough of the belly of a ship, crawling around in cramped quarters and changing shifts every couple of weeks. I like to work days and be able to see other people and stretch out if I feel like it. It's noisy working on board a ship, because you're surrounded by metal. We did everything from changing light bulbs to maintaining the huge electric generators. It was a good experience, and I feel we got the best training.

"It's important to have coworkers who are as safety-conscious as I am. If the people around you are careless, their actions could cause you to fall, or be hit by something else that's falling. And there's always the possibility of electrical shock.

"When I took my journeyworker's exam, there were people across the hall taking the apprenticeship test. The union and contractors have all these new programs to recruit women and minorities. Our company is supposed to get two new apprentices the first of the year. It will be interesting to see who they are. It was a pretty mixed group of folks."

Phil

Phil is a journeyworker electrician whose fifteen years in the trade include five years as the licensed operator of a contracting business. As both job estimator and office manager he supervises a staff of ten electricians plus a secretary-bookkeeper.

Blueprints and electric supply catalogs litter his desk, for work is under way to bid the electrical contract for a municipal park project. The plans call for two buildings, a bathhouse near the swimming pool, and a large pavilion, plus lighted tennis courts, playground, and parking area. The equipment list also calls for an electrically powered filtration system for the swimming pool.

"Our job, and I sure hope we get it, will be to supply and install all these elements of the project," Phil explains, with a sweeping gesture that takes in the blueprints.

"We need two or three big projects like this each year to keep things on an even keel. Last year we got the bid for nearly $1 million in renovations to the local high school—new classroom lighting, a new public address system, and cable TV to most of the classrooms. The rest of the year, we're okay with service calls and maintenance, but we really need a project of some scope to keep everybody busy and the business growing."

As he speaks, a computer printer on a nearby table is spitting out long green and white sheets. They are material orders and inventory lists.

"It's kind of hectic around here today, but all this means work for months ahead for the crew; that is, if we get a couple of these jobs."

Phil began his career as an apprentice, followed by

five years as a journeyworker. After that came assignments as job superintendent and electrical cost estimator for several other contracting firms. He started his own business three years ago.

"Every job I took was calculated to help me learn more about the business side of electrical work. I always knew I wanted to be my own boss. This company is small, but it's growing. The next phase is to hire someone to do the estimating work, which will cut my workday from twelve hours in the busy season to something more manageable. I'll still meet with the sales reps and general contractors, but my goal is to hang up my hard hat for good some day and just run things from the office.

"Planning projects, investigating new materials and methods of doing a job, learning to motivate my employees—those are the things I enjoy most now."

Phil is also interested in the apprentices coming into the trade. "Seeing people develop into skilled workers with safety and pride in their work—that's a great thing to witness."

KIM

Moving to the world of the utility company worker, we focus on meter installer who learned the work in on-the-job training, or OJT.

A high school graduate who had been laid off from an unskilled warehouse job, Kim was referred by the unemployment office to a six-month training program in basic electricity.

"I really didn't know where the classroom training would lead me, but after being laid off three times in two years, I knew I had to get some skills. We had some remedial stuff at first, just to freshen up on math and following directions. Right from the start, though, we were learning about electricity and how it is conducted. There was a lot of safety instruction, too, and job-search skills. They taught us about role models and how to make decisions, and they took us on field trips to see people from earlier classes and the kinds of jobs they had gotten.

"By the time I was called for an interview for this job, I'd been through three or four practice sessions in class. I felt more relaxed and confident than ever. I guess I came across as having some potential, 'cause here I am."

Kim's job title is meter installer, and she works for a major East Coast power company. She worked as an installer's helper at first. Once she proved she could handle the various details, the company sent her out on solo assignments.

"I install electric meters, mostly, but sometimes I inspect them and take out the ones that don't work. I might work at private homes or apartment buildings, a store, or even a factory, all in the same day. I have a company van to drive from one assignment to the next. On occasion, I have to climb into a manhole to work."

Sometimes Kim's work takes her to buildings under construction. "The first step is to get the work orders from the service building. The details are all printed out there, along with the locations. The work requires some lifting. Transformer cabinets weigh thirty pounds or more, and sometimes you need to move them. The meters weigh about ten pounds each. If there's no elevator, you have to carry them up stairs."

Glancing at her hands, she notes, "They can be a

manicurist's nightmare. Sometimes we work with heavy wire that cuts up your hands. There are times when you get pretty dirty, and my head sweats under the hard hat, which bothers me some. Still, the paycheck makes it worthwhile, and this is a secure job with good benefits."

The job also has a measure of independence that Kim values. "There's a radio in the van, and I check in with the dispatch office between assignments, but other than that, I'm trusted to get the job done. There's no one looking over my shoulder all the time, the way factories operate. That makes me proud, and to tell the truth, I've never felt that way about a job until this one."

While these "word pictures" cannot convey the reality of a career, they are based on real-life experiences of electricians and electrical workers. These workers are among the highest blue-collar workers in America, enjoying stable employment and the respect of fellow workers.

As America moves closer to becoming a service-based economy, there will be fewer opportunities for workers to measure individual accomplishments. Not so for electricians, who literally light and power the world around them.

11

Jobs of the Future

In a residential neighborhood in the gently rolling Massachusetts hills stands an attractive family home that marks a milestone in housing materials research. More than 45,000 pounds of plastic was used in building this twenty-first-century house, including parts of the electrical system—not just any plastic, however, but space-age polymers and thermoplastics. The house has sophisticated electrical wiring in high-tech panels and raceways, and it is just one example of what lies ahead for construction electricians.

Circling the perimeter of the rooms is a wire management system that looks like traditional baseboard trim but conceals both electric current and communications cables side by side. The device keeps the wiring accessible for upgrading or service, yet allows flexibility in locating electric outlets and telephone jacks.

The futuristic home also boasts windows that switch electrically from see-through clarity to a highly opaque state for privacy at the touch of a switch. In the entertainment room, there is a 43-inch flat-screen television and ceiling-mounted audio speakers, part of a whole-house audio design hardwired into the electric system.

Technology and the need to meet new challenges

will bring new tools, products, and techniques to the electrical industry as it enters the twenty-first century. No matter which type of electrical work you choose, it will include the need for continuing education and a flexible attitude toward the work you do.

Electricians and electrical workers will deal with increased use of prefabricated components and modular construction. The construction and electric utility industries will see changes related to improved energy efficiency. In all areas, especially in communications and power generation, the trend is increasingly toward automated systems and equipment.

As you have already learned, lineworkers in the cable television and telephone industries will be most affected by the decline in hirings brought about by fiber optics. In all other phases of the electrical industry, job growth is anticipated. The construction industry will be hiring all the skilled electricians it can get, and in keeping with the laws of supply and demand, wages will rise. There will be dramatic growth in job opportunities for female and minority workers, with active recruiting programs to train them for jobs.

As you may have guessed, the electricians who wired the house described above encountered some "firsts" in their careers, but they will definitely encounter new firsts as the building industry moves into the next century. The General Electric Company has begun marketing the innovative products used in this home. All are designed to improve the quality of life. Electricians who were long concerned with safety and convenience are taking this third concept into account when planning new projects.

Talk about convenience—in the "mud room," the

project electricians installed heater cables that pull out from a wall-mounted storage cubicle, ready to dry wet shoes or boots. Nearby is a sophisticated "environment module," wired to control space and water heating, air conditioning, and dehumidifying equipment for this house. Each of these systems is a module. If any module fails to function properly, the homeowner can diagnose the problem with a handheld probe.

The master bathroom has an electronic panel as well, designed to monitor and record such vital signs of human health as pulse and blood pressure, transmitting the data electronically to a doctor's office. A tiny video camera built into a speculum allows the doctor to make an electronic house call. Say you have a sore throat. Just dial the doctor's phone number, open wide, and say "Ahh."

PROBLEM-SOLVING

Solving environmental problems, advancing space exploration, alleviating transportation problems, and curbing the waste of precious natural resources will all play a vital role in increasing our standard of living and thus shape the way we live and work. These important activities will also create new jobs and bring about change in existing jobs. Much of today's technological progress is the result of success in the related fields of electricity and electronics, and it is likely that these industries will continue to grow, reflecting discoveries and inventions appearing just now on the horizon.

The combination of electrical and electronics skills is in particular demand in the space program. Although job opportunities may be shrinking in these areas as a result of governmental budget cuts, many electrical workers are employed in aerospace and defense industries. Job titles listed in the "growth category" include instrumentation mechanics, test technicians, and control specialists.

Two new emerging technical fields are fiber optics and robotics. In 1966, scientists working for International Telephone and Telegraph (ITT) made lightwave communication practical. As the technology was refined, new methods and products evolved into the present day, when fiber optic cable systems exert a tremendous influence on communication systems.

The development of high-definition television (HDTV) is sure to bring many innovations to consumer electronics, as will advances in computer technology and robotics. Because each relies on intricate wiring and control systems, expert electricians will be needed to build, maintain, and repair this equipment.

In the decade ahead, the job structure in operation for skilled electrical workers is not expected to change drastically, although new applications for existing technology will continue to be introduced. Energy-efficient, cost-effective equipment and the methods of installation for that equipment will be introduced. For the industry overall, new products, equipment, and machinery will continue to emerge. Electrical workers must be flexible, willing to learn and to change with the times, if they hope to keep pace with their trade.

Electricians will be involved with solving many of the problems that lie ahead for the world, including environmental issues and space exploration and travel. Here on earth, primary goals include alleviating mass transit problems in major cities and eliminating uneconomical uses of natural resources. The electrical industry and its cousin, the electronics industry, are likely candidates to play vital roles as the world enters a new era.

SMART HOMES REQUIRE ELECTRICIANS

To illustrate how these changes will affect the homes we live in, let's consider the House of the Future, created by Southern California Education's Customer Technology Application Center (CTAC). The 1,000-square-foot home opened in 1990 and serves as a model for centers now being built across the United States. The house reflects the electric utility industry's concerns about energy efficiency and the environment. Visitors can see demonstrations of high-efficiency heat pumps, electronic meters, water conservation options, electric vehicles, and home automation systems.

For instance, the kitchen of the future boasts a collection of computer-assisted appliances. Picture the dishwasher, loaded and ready to go. The homeowner pushes the start button, but instead of initiating the wash cycle, the dishwasher electronically contacts the local electric utility, determining when the most economical rates are in effect, and waits until then to wash the dishes. The oven and the laundry equipment are similarly equipped with "brainpower," allowing the occupants of this futuristic home to "call" the appliance via touch-tone telephone and signal its start-up. Inexpensive and plug-in devices for the coffeemaker, lamps, and even the drapery hardware permit long-distance operation via tone signals. Creating, installing, maintaining, and repairing these devices will still require a qualified electrician.

SOLVING URBAN PROBLEMS

As for safeguarding the environment, we are likely to see the development of new and better electronic precipitators capable of removing larger quantities of pollutants from the atmosphere. There will be more elaborate water purification systems that will use sophisticated electrical and electronic components, and we will enter into a new era of electrified sanitation centers that deal with disposal of solid wastes.

Many of today's traffic-clogged cities depend on inadequate, outmoded public transportation systems and highway networks. Both contribute to pollution and erode the quality of life. Safe, clean, electrically powered rapid transit systems for cities may well be the answer to this problem as well. And you guessed it: Each of these projects will generate job opportunities for electrical workers.

The twenty-first century will be bright indeed for skilled workers who know how to harness and direct the immense energy source we know as electricity.

Glossary

agreement, collective bargaining A contract negotiated between a union and an employer to cover workers' wages, hours, fringe benefits, and working conditions.

apprentice A person, usually between eighteen and twenty-four years of age, who learns a trade through a combination of on-the-job training and classroom instruction.

blueprint A drawing, reproduced by a photographic process, depicting an architect's or designer's construction drawing.

building system System in which materials, prefabricated units, and labor are scheduled for use to maximize efficiency. Also called systems construction.

building trades Skilled trades represented in the construction industry, including bricklayers, carpenters, electricians, masons, painters, and plumbers.

collective bargaining A method of determining employment conditions by negotiating between the employer's representatives and employees' union representatives.

compensation A package of wages and benefits, both current and deferred (as with vacation time or retirement pensions), that workers receive in exchange for their work.

contractor An individual or a company engaged to do a specific job under conditions and prices spelled out in a legal agreement called a contract.

craftsperson An artisan whose work or occupation requires particular training and practice to achieve a high level of expertise.

estimator A worker who calculates the materials, labor, and general costs required to accomplish a job.

fabrication The act of constructing an item from standardized parts; components that arrive at a construction site ready for installation are said to be fabricated.

fiber optics A system that transmits light signals through glass fibers. Fiber optic cable, flexible and lightweight, is capable of carrying more than 50,000 telephone conversations simultaneously, yet is no larger in diameter than your finger.

finish work Final, sometimes decorative, work, usually requiring expert care and skill, done at the end of construction.

foreperson Leader of a work crew, usually specially trained or more experienced.

fringe benefits Benefits in addition to the basic wage rate, such as health insurance coverage and paid vacation days.

inspector Person who examines a project or job element to guarantee that standards of safety or quality are being met.

journeyworker A worker who has completed apprenticeship training and is prepared for full responsibility and earning potential in a construction job.

laser (**l**ight **a**mplification by **s**timulated **e**mission of **r**adiation) A device that uses the natural oscillations of atoms or molecules between energy levels for generating coherent electromagnetic radiation in the ultraviolet, visible, or infrared regions of the spectrum.

layoff An interruption in employment, usually related to a construction slowdown or a temporary halt at a work site because of inclement weather.

layout work Reading of blueprints and specifications and translating that information into specific instructions for workers.

manual skills Those performed by hand, requiring the use of skill or energy.

modular Work involving precision or standardized parts made for fast and efficient assembly on-site.

on-the-job training (OJT) Paid employment that combines work with learning.

overtime Time spent on the job that exceeds the basic workday or workweek as defined by law; premium rates of pay apply to such hours.

robotics Technology dealing with the design, construction, and operation of robots in automation.

service call Repair work performed on the customer's premises, rather than in shop or factory.

shift A person's daily work schedule in a firm that operates continuously over a twenty-four-hour period. Shifts are customarily 8 AM to 4 PM, 4 PM to midnight, midnight to 8 AM.

shop School course that emphasizes the use of tools in a trade or craft.

standard A general model of safety, quality, or performance to which other work is compared.

subcontractor A person or company that agrees to perform specific skilled work on a building being erected by a general contractor.

superintendent A supervisor who directs the work of forepersons and their crews at a construction site.

troubleshooting The process of testing and/or diagnosing malfunctions in electrical circuits, appliances, machinery, or their equipment.

wage scale Listing of wages from highest to lowest, comparing compensation for specific job categories.

Appendix

The organizations listed below may be of interest in helping to plan your career as an electrician or electrical worker. Those with a site on the World Wide Web are marked with the corresponding Internet address. Where specific apprenticeship or training programs are available, that information is noted.

Electrical Apparatus Service Association (EASA)
1311 Baur Boulevard
St. Louis, MO 63132
(314) 993-1269
Web site: http://www.electricnet.com

An active trade association in the electrical industry, this is an organization of independent firms engaged in the repair, maintenance, and sale of electric motors, transformers, generators, controls, and associated equipment.

Women's International Network of Utility Professions (formerly Electrical Women's Round Table)

P.O. Box 335
Whites Creek, TN 37189
(615) 876-5444
Web site: http://www.winup.com
E-mail: winup@aol.com

Founded in 1927, this national organization represents women holding positions connected with the electrical industry or allied fields including communicator, educator, information specialist, and researcher. Objectives are to promote knowledge and expertise among members, to increase recognition and encourage upward mobility of women in the electrical industry, and to advance consumer education.

Electrification Council
701 Pennsylvania Avenue NW
Washington, DC 20004
(202) 508-5901
Web site: http://www.ei.org/tec
E-mail: tec@eei.org

Comprises five trade associations and fifteen manufacturers that work to support the efficient use of electrical energy. The council provides training courses for commerce and industry on such topics as energy management, industrial and commercial lighting, and power distribution, motors, and motor controls.

Independent Electrical Contractors (IEC)
2010-A Eisenhower Avenue
Alexandria, VA 22314
(800) 456-3424
(703) 549-7351
Web site: http://ieci.org

With nearly 3,000 members, the IEC represents independent electrical contractors, small and large, primarily open shop. It sponsors electrical apprenticeship programs, and, among other responsibilities, it represents independent electrical contractors to the National Electrical Code panel.

International Brotherhood of Electrical Workers (IBEW)
1125 15th Street NW
Washington, DC 20005
(202) 833-7000
Web site: http://www.ibew.org

A union of nearly one million electrical workers, the IBEW cooperates with employers and government in sponsoring apprenticeship training. It publishes the monthly magazine *IBEW Journal.*

National Electrical Contractors Association (NECA)
3 Bethesda Metro Center
Suite 1100
Bethesda, MD 20814
(301) 657-3110
Web site: http://www.necanet.org
E-mail: webmaster@necanet.org

For more than eighty years, NECA has worked to improve the quality of electrical service. It works with the IBEW to cosponsor the National Joint Apprenticeship and Training Committee for the Electrical Contracting Industry, which sets standards and works to ensure an adequate supply of trained electrical workers. NECA also offers services to its more than 6,000 small-business members.

National Association of Women in Construction (NAWIC)
327 South Adams Street
Fort Worth, TX 76104
(817) 877-5551
Web site: http://www.nawic.org
E-mail: nawic@onramp.net

Founded in 1954, NAWIC currently has nearly 9,000 members and represents professional women in the construction industry, including electrical fields. It educates members in new construction techniques, awards national and local scholarships, and assists local groups with employment services and career day workshops.

National Fire Protection Association (NFPA)
1 Batterymarch Park
P.O. Box 9101
Quincy, MA 02269-9101
(617) 770-3000

Founded in 1986, the NFPA is a nonprofit, voluntary organization with more than 31,500 members world-wide. Its purpose is to safeguard people and buildings and their contents from hazards arising from use of electricity for light, heat, power, radio, and signaling.

The electrical division allows members to become better informed on electrical safety. The NFPA has contributed to development of the National Electrical Code and other NFPA electrical standards.

Wider Opportunities for Women, Inc. (WOW)
815 15th Street
Suite 916
Washington, DC 20005
(202) 638-3143
Web site: http://www.w-o-w.org
E-mail: info@w-o-w.org

National Urban League's Labor Education Advancement Program (LEAP)
112 Wall Street, 8th Floor
New York, NY 10005
(212) 558-5300

Job Corps
United States Department of Labor
E7A
Office of Job Corps
3535 Market Street
Room 12220
Philadelphia, PA 19104
(215) 596-6301
Web site: http://www.jobcorps.org

The Home Builders Institute
1090 Vermont Avenue
Suite 600
Washington, DC 20005
(800) 795-7955
Web site: http://www.hbi.org

YWCA of the USA
Empire State Building
Suite 301
350 Fifth Avenue
New York, NY 10118
(212) 273-7800
Web site: http://www.ywca.org

For Further Reading

Adamczyk, Peter, Paul F. Law, and Andy Burton. *Electricity and Magnetism.* Tulsa, OK: EDC Publications, 1994.

Alerich, Walter N. *Electrical Construction Wiring.* Albany, NY: American Technical Publications, Inc., 1993.

Electrician and Electrician's Helper. 6th ed. Englewood Cliffs, NJ: Prentice-Hall, 1983.

Fowler, Richard J. *Electricity: Principles and Applications.* Westerville, OH: Glencoe/McGraw-Hill, 1993.

Lewis, Maurice. *Electrical Installation Guide.* Woburn, MA: Butterworth-Heinemann, 1997.

Lytle, Elizabeth Stewart. *Careers in the Construction Trades.* New York: Rosen Publishing Group, 1995.

Neufeld, Rose. *Exploring Nontraditional Jobs for Women.* Rev. ed. New York: The Rosen Publishing Group, Inc., 1989.

Nye, David E. *Electrifying America: Social Meanings of a New Technology.* Cambridge, MA: MIT Printing, 1992.

Occupational Outlook Handbook. Washington, DC: Bureau of Labor Statistics, U.S. Department of Labor.

Ricci, Larry J. *High-Paying Blue-Collar Jobs for Women: A Comprehensive Guide.* New York: Ballantine Books, 1981.

Rosenberg, Paul. *Electrician's Book of Trade Secrets.* Fairfax, VA: SER Publications, 1987.

Shertzer, Bruce. *Career Planning: Freedom to Choose.* 3rd ed. Boston: Houghton-Mifflin, 1990.

Skipp, Ray. *Electrical and Electronics Wiring Guide.* Woburn, MA: Butterworth-Heinemann, 1998.

Smith, Russell E. *Electricity for Refrigeration, Heating, and Air Conditioning.* Albany, NY: Delmar Publishers, 1997.

Spence, William P. *Construction: Industry and Careers.* Englewood Cliffs, NJ: Prentice-Hall, 1990.

Starr, William. *Electric Wiring and Design: A Practical Approach.* Englewood Cliffs, NJ: Prentice-Hall, 1983.

Sumichrast, Michael. *Opportunities in Building Construction Trades.* Lincoln, IL: National Textbook Co., 1985.

Traister, John E. *Electrician's Exam Preparation Guide.* Carlsbad, CA: Craftsman Books, 1995.

U.S. Navy. *Basic Electricity.* Dover Publications, 1975.

Valkenburgh, Nooger Van. *Basic Electricity.* Indianapolis, IN: Prompt Publications, 1993.

Whitson, Gene. *Handbook of Electrical Construction Tools and Materials.* New York: The McGraw-Hill Companies, 1996.

Wood, Robert. *Opportunities in the Electrical Trades.* Lincolnwood, IL: National Textbook Co., 1991.

Index

A

aerial basket, 29, 35
apprenticeship, 4, 5, 12, 38, 42–43, 46–47, 50, 59–79, 118
  agreement, 62–63

B

Bell, Alexander Graham, 10
blueprints, 7, 21, 23, 31, 61, 71, 77, 84, 105, 109, 124
boss, being your own, 108–113

C

cable splicer, 6, 28, 30–31, 45, 60, 80
  apprentice program for, 81–82
classroom instruction, 59, 63, 70–79, 81
communications, 2, 7, 36, 41, 104
Communications Workers of America, 45, 87
conduit, bending, 18, 30, 56, 63
construction electrician, 2, 5, 6, 19, 20–21, 41, 83, 88, 114–115
  apprenticeship program for, 63–64
construction supervisor, electrical, 106
contractor, electrical, 21, 37, 71, 102–103, 109, 118
controllers, 22, 23, 24
control specialist, 101–102, 130
Council of Industrial Relations (CIP), 40
Craft Skills Program, 48

D

dangers, electrical, 8, 12, 19, 29, 39, 102, 111
degrees
  associate in electrical occupations, 83
  associate in electronics technology, 84–85
dishwashers, 6, 72, 131
dues, union, 42–43

E

Edison, Thomas Alva, 10
Electrical Wiremen and Linemen's Union, 39
electrical workers, days of, 114–126
electrician, history of, 9–18
electricity, 2, 4, 7, 9–11, 104, 129
electric signs, 7, 12, 19, 20, 33–35
  apprentice program for, 70

electric utility company, 2, 3, 6, 19, 60
electronics, 2, 5, 7, 12, 61, 63, 78, 84, 101, 104, 129
  journeyworker, 103–104
engineering, electrical, 12–13, 104
environment, safeguarding, 132
estimator
  cost, 77, 124
  job, 103, 105, 110, 123–124
  material, 71, 110
exam
  aptitude, 59–60
  journeyworker, 50, 122
  physical, 59, 88
  union admission, 42

F
fiber optics, 5, 7, 83, 85, 128, 130
fringe benefits, 3, 21, 38, 44–45, 57, 63

G
generator, 11, 22, 24, 25, 103
ground fault circuit interrupter (GFCI), 13–17, 72, 76

H
heating/air conditioning, 6, 20
Home Builders Institute, 48
hydraulics, 101, 104

I
independent electrician, 44, 51, 57, 114
Individual Retirement Account (IRA), 44
initiation fee, union, 42
inspector, electrical, 17, 52, 55, 75, 77–78, 83, 102
instrumentation technician, 19, 101, 104–105, 130
insurance, health, 3, 21, 38, 45, 57
International Brotherhood of Electrical Workers (IBEW), 36, 37–44, 78, 81, 115, 119
  in Canada, 41, 79
International Union of Electronic, Electrical, Salaried, Machine, and Furniture Workers, 45
interview, personal, 59, 61, 88, 89

J
job
  application, 89, 93
  interview, 96–100, 125
  preparing for, 87–100
Job Corps, 47–48, 82–83
jobs
  bidding for, 88, 109, 110, 123
  of the future, 127–132

L
Labor Education Advancement Program (LEAP), 49
labor union, role of, 37–45, 60
Labor, U.S. Department of, 5, 49, 60, 78
lasers, 2, 5, 7, 83, 84–85
layoffs, seasonal, 88, 121
license
  journeyworker, 50
  master's, 50, 56
line crew chief, 106
line worker, 6, 28–29, 45, 60, 80, 128
  apprentice program for, 67–68, 81

M
maintenance electrician, 2, 3, 5, 6–7, 21–22, 47, 83–84, 104, 114
  apprentice program for, 64–65

manufacturing, electrical, 7, 13, 19, 36, 41, 104
marine electrician, 7, 20, 24–26, 47
  apprentice program for, 66, 67
marketing, 110–111
materials, 17, 71, 77, 84, 105, 119
McNulty, Frank J., 40
mechanics, 101
mentor, 112
meter installer, 124–125
military service, 26, 60, 83, 104–105, 121–122
Miller, Henry, 40
minorities
  in apprenticeship, 60
  in electrical industries, 4, 46–58, 128
motors, electric, 7, 12, 20, 23–24
  apprentice program for, 65–66

N
National Association of Home Builders, 46, 48, 78, 110
National Association of Women in Construction (NAWIC), 49, 112
National Electrical Code, 8, 23, 31, 34, 52, 74–75, 84, 102, 121
National Electrical Contractors Association (NECA), 43–44, 78, 103
National Joint Apprentice Training, 62
National Urban League, 49

P
pay, electricians', 3, 21
pensions, 3, 38, 41, 43–44
personnel, 111–112
pneumatics, 101, 104
Power Line Pro, 81–82
power plant maintenance electrician, 28–32, 33, 47
  apprentice program for, 68–70

Q
qualifications, apprenticeship, 60–62
questions, asking in job interview, 96–97

R
résumé, 89
  samples, 90–93
robotics, 2, 5, 7, 13, 130

S
safety rules/standards, 39, 53, 75, 84, 102, 111, 117, 122, 125
service company, electrical, 19, 110, 119–121
Service Corps of Retired Executives (SCORE), 112
shift work, 7, 22, 28, 95
shop supervisor, 10
Smart House, 78, 131
solar power, 18
space exploration, 2, 5, 129
specializing, 101–107
Specific Aptitude Test Battery, 59–60
supervisor, 101, 102, 103, 106
supplemental unemployment benefits (SUB), 21, 43, 88
support network, 112–113

T
teaching, 13, 106–107
television, high-definition, 130
test
  equipment, 7, 22, 24, 72, 84, 102, 103, 104
  technicians, 130

tools, 23, 24, 32, 35, 61, 63, 71, 82–83, 84, 105, 119
trade school, 60
training
on-the-job (OJT), 4, 33, 59, 60, 63, 71, 81, 88, 124
other routes to, 80–86
transformers, 24, 29, 30, 31, 34
troubleshooting, 22, 28, 29–30, 84, 102, 103

U
Underwriters Laboratories (U.L.), 15, 75
utility company, 11–12, 26–28, 41, 102
apprenticeships, 81–82
jobs with, 80–82, 124–125

V
vacations, 3, 41, 44, 45

W
weather, 3, 26, 29, 56, 61, 85
Wider Opportunities for Women (WOW), 48
wiring, residential, 13–18, 20, 51–52, 73–74
woman electrician, story of a, 49–58
women
in apprenticeships, 60
in electrical industries, 4, 12, 46–58, 112, 118–119, 128
Women's Bureau, 49
workday, length of, 38–39

Y
Young Women's Christian Association (YWCA), 48–49